The Contest Problem Book V

American High School Mathematics Examinations
and
American Invitational Mathematics Examinations
1983–1988

Problems and solutions compiled and augmented by

George Berzsenyi
Rose-Hulman Institute of Technology

Stephen B Maurer
Swarthmore College

D0922477

38

THE MATHEMATICAL ASSOCIATION OF AMERICA

NEW MATHEMATICAL LIBRARY
published by
The Mathematical Association of America

The New Mathematical Library (NML) was begun in 1961 by the School Mathematics Study Group to make available to high school students short expository books on various topics not usually covered in the high school syllabus. In three decades the NML has matured into a steadily growing series of over thirty titles of interest not only to the originally intended audience, but to college students and teachers at all levels. Previously published by Random House and L. W. Singer, the NML became a publication series of the Mathematical Association of America (MAA) in 1975. Under the auspices of the MAA the NML will continue to grow and will remain dedicated to its original and expanded purposes.

©1997 by the Mathematical Association of America
ISBN 0-88385-640-9
Library of Congress Catalog Number 97-70508
Printed in the United States of America
Current Printing
10 9 8 7 6 5 4 3 2 1

The Contest
Problem Book V

American High School Mathematics Examinations
and
American Invitational Mathematics Examinations
1983–1988

NEW MATHEMATICAL LIBRARY

Other titles in preparation.

Contents

Preface

This is the fifth book of problems from the American High School Mathematics Examination (AHSME), covering the six examinations from 1983–88. It is also the first book of problems for the follow-up American Invitational Mathematics Examination (AIME), which began in 1983. These are two of the four examinations in the American Mathematics Competitions (AMC); the others are the USA Mathematical Olympiad (USAMO) and the American Junior High School Mathematics Examination (AJHSME).

Each AHSME consists of 30 multiple-choice questions, and each AIME consists of 15 questions, with each answer an integer from 0 to 999. Both examinations cover precalculus material. Contestants have 90 minutes for the AHSME, 3 hours for the AIME (2.5 hours in 1983–85). During the period covered by this book, about 400,000 students took the AHSME each Spring and 1000–4000 were invited to the AIME (based on AHSME score). On both exams, problems are roughly in increasing order of difficulty, with the AIME questions, on average, much harder. The AHSME is one of the largest mathematics competitions in the world, and all the AMC exams are known and respected worldwide.

Why Buy this Book?

The immediate reason is to practice in order to do better on future offerings of these competitions. And there is no doubt that practice helps. The AHSME is a hard exam: typically the best score in a school will be less than 100 points out of 150 possible, so there is plenty of room for improvement.

However, the fundamental goal of the mathematics community in providing these exams is to pique interest in mathematics and develop talent. Problems are at the heart of mathematics, and experience greatly sharpens problem-solving skills. We believe that some of the problems on the AHSME and AIME are intriguing in themselves, and/or they bring out important mathematical points. Thus, one doesn't ever have to participate in the AHSME or AIME contests to gain something from this book – older readers aren't eligible anyway! Indeed, working these problems from this book you need not stick to the artificial time limits that contests impose. Often the best solutions, and the deepest learning, come when there is time to reflect.

While supplies last (many years), each exam and its solution pamphlet are available individually. This book offers the convenience and lower cost of bundling – and much more:

1. Additional solutions. Although most solutions in this book are the same as provided in the complete solution manual made available right after each exam is given, there have been some editorial improvements in them and some additional solutions have been added. Some of these additional solutions weren't in the solution pamphlets due to space limits, some are solutions mailed to us by students and teachers, and some were provided by members of the MAA Publications committee when they reviewed our manuscript.

2. An index. If a reader wishes to work on specific types of problems, this index makes it possible.

3. Pointers to other material. We have added some further references to related problems and to articles and books that expand upon ideas in some of the problems.

4. "Dropped Problems". In order to get good exams, one must begin with many more problems than will fit. Invariably, some very good problems don't make the final cut. We provide two sets of dropped problems (heretofore never revealed!), one for the AHSME and one for the AIME.

5. Statistical information. The AHSME is a multiple-choice exam. We provide frequency-response tables that show how popular each answer was for each question. The introduction to that section explains how one can use these tables to calibrate one's work and identify important mathematical errors.

How to Use This Book

You can learn from this book through both the problems you get right and the problems you get wrong. After doing a problem (or a set of problems – but you don't need to stick to the time limits), first look at the answer key. If you got the right answer, then look at our solution(s) to compare with yours. If you are fortunate, our solution will be very different from yours and will thus introduce you to an alternative approach. If you got the wrong answer, go back to the problem before looking at the solution. Just knowing that your answer was wrong (or sometimes knowing the right answer) will jog you into seeing your error and lead you to a correct solution. If not, then look at the solution. Ask yourself: What is the key idea in this solution that I missed (or misunderstood)?

Another good thing to do with any problem is to change the hypotheses and see how the conclusions change. More generally, see if you can use ideas from this book to make up and solve new problems at the right challenge level for you and your friends. Problem *posing* can be as good a learning experience as problem solving.

We hope this book will be used not just by individuals, but also by math clubs or groups of students working together in class or elsewhere. Sharing ideas – from initial false starts through to a joint solution – is sometimes the most exciting way to do mathematics.

Do not be discouraged if there are many problems you cannot solve. These are hard exams and are meant to challenge almost everybody. Keep in mind that during 1983–88 at most a few thousand students obtained an Honor Roll score on the AHSME (100 out of 150). The fewest Honor Roll students was 624 in 1984, the most 8050 in 1988. Also, during these 6 years only 7 students obtained a perfect AHSME score of 150. The AIME is even more difficult. On average, the extremely able students who took it scored about 5 out of 15, and only 11 students obtained a perfect 15 during these years. If you find, for instance, that the last 10 problems on the 1983 AHSME are beyond you, then on later AHSMEs concentrate on problems just before and after this cutoff point. Stretching yourself just beyond what you can do easily brings about the best growth.

In particular, younger students especially should not be concerned that they cannot do all the problems. They haven't even studied some of the topics covered yet. If a problem treats a topic you don't know, just skip it.

History and Changes

The AHSME was first given in 1950, under the name Annual High School Contest; it was then sponsored by the New York Metropolitan Section of the Mathematical Association of America (MAA) and was offered in that region only. In 1957 it became a national competition, cosponsored by the MAA and the Society of Actuaries. By 1982 it was sponsored by five organizations: The MAA, the Society of Actuaries, Mu Alpha Theta, the National

Council of Teachers of Mathematics, and the Casualty Actuarial Society. Also, it was given to over 400,000 students in American and Canadian schools in North America and elsewhere, and translations were in use by many other foreign schools.

During 1983–88 there were a number of specific changes and some gradual evolutions.

Names

Unfortunately, the AHSME had been known by a number of names over the years, all in use simultaneously, none of them fully descriptive, and none of them short. So in 1983 "American High School Mathematics Examination" and the acronym AHSME were introduced, with emphasis on the latter. This acronym seems to have caught on.

For similar reasons, and because a middle-school examination was now involved, the examination program as a whole became the American Mathematics Competitions (AMC).

More sponsors

Because of growing concern about the quality of pre-college mathematics education, many organizations that had not been directly involved at this level decided that it was important to become involved. As one part of this change, the AMC received two new sponsors: The American Statistical Association in 1985, The American Mathematical Association of Two-Year Colleges in 1986. Since 1988, further sponsors have come on board: The American Mathematical Society in 1989, and the American Society of Pension Actuaries in 1995.

More exams

The AIME commenced in 1983. A later section presents the reasons for introducing the AIME. A junior-high exam, modeled after the AHSME, commenced in December, 1985: The AJHSME (American Junior High School Mathematics Examination).

AHSME scoring system

The scoring method changed once during this period. From 1978 to 1985 the formula was $30 + 4R - W$, where R is the number of correct answers and W is the number wrong. From 1986 on the formula has been $5R + 2B$, where B is the number you *don't answer* (leave blank). There are 30 questions, so note that the top score is 150 either way. To learn the reasons for this change, see the section "AHSME Scoring" below.

AHSME difficulty level

For many years there had been a debate within the AMC about making the AHSME easier. On the one hand it was the capstone challenge mathematics examination for North American high school students, harder than various state and local exams and with a long tradition. On the other hand, efforts to keep many students and their teachers from getting discouraged, by having them set realistic goals, were not succeeding. The introduction of the AIME provided a context in which a consensus was reached to make the AHSME easier.

Progress was made on this goal during 1983–88. The number of Honor Roll students in these years was, by year, 1823, 624, 931, 2848, 4221, and 8050. However, setting the difficulty level of a competitive exam in advance is not so easy, and even judging the difficulty after the fact is not clear-cut. As for after-the-fact judgments, a different set of students takes the exam each year, so even if the average score was the same in two consecutive years, that suggests only that the exams were equally hard. Furthermore, the scoring changed in 1986, and the change in Honor Roll cutoff may not have compensated exactly.

As for before-the-fact judgments, the AMC exams are not standardized tests, like the SATs. The College Board uses some problems over again (so called "equators"), which allows them to figure out how to set scores so they have the same meaning each year. They can have equators because not all of their tests are made public; sometimes all the test problems are collected and not published later. However, the AMC immediately makes all its problems and solutions publicly available – our main goal is to promote mathematical learning. While we (meaning all the AMC test writers) try to judge the difficulty of our tests in advance, by averaging our opinions, this is merely a rough guide and our judgments are often off.

Changes in coverage

Over time the style and emphases in mathematical problems change. When the AHSME started, it consisted almost exclusively of classical geometry and algebraic manipulation. (Indeed, it seems that trigonometry did not appear explicitly until 1972, although alternative solutions to some earlier problems used trigonometry.)

By 1982 AHSME coverage had become much broader. The main changes during 1983–88 were the following.

Addition of statistics. Probability and statistics are now recognized as important topics for citizens to learn about through their secondary mathematics courses. For this reason, and because the American Statistical Association became a sponsor, occasional problems on statistical concepts were included, e.g., averages, modes, best fit. Also, more problems about probability appeared.

Algorithms. Calculators, computers, and methods of computation have become more important, so AMC exams have paid more attention to related mathematics. For instance, there were more problems about recurrences (sequences where each term is defined by using earlier terms) and problems about the end result of repeated calculations.

Applications. We wanted to stress applications more. But there is a difficulty. Any realistic application is complicated to describe, and at least half the battle is coming up with (and justifying) a reasonable model. This makes it very difficult to write an applications problem that is simple and clear enough for the AHSME or AIME. The leading applied math contest, the COMAP Modeling Contest, is a group contest where students have a whole weekend and must write an essay in which, among other things, they can explain and justify their interpretation of the problem.

Computation. More generally, there has been a trend in mathematics education to downplay mechanical computation and emphasize concepts. The AMC exams involve much less computation now than in the early years, when students had only 80 minutes for 50 questions – speed and precision were at a premium then! Throughout 1983–88 the AHSME had 90 minutes for 30 questions, and for the AIME there was much more time per problem (which increased when the total time was extended from 2.5 hours to 3 hours). However, there are deliberately still a few problems where, say, algebraic prowess is still needed.

Allowed Aids. What is interesting here is that nothing changed. Computers, calculators and slide rules remained forbidden, graph paper, rulers, compasses, protractors were allowed. There was some discussion about permitting calculators, but they were first allowed much later – 1994 – and only on the AHSME, not the AIME. There are a few problems in this book where a hand calculator makes the problem much easier; such problems could not be used on the AHSME today.

Distractor change

"Distractor" is test-maker language for wrong answer. The use of this word suggests that the answers are there to "trick" you. During the period covered by this volume the AMC test writers decided to limit the ways in which distractors could be tricky.

Suppose you are asked to find the diameter of a sphere meeting some condition. Suppose further that you do a lot of good thinking and hard work to prove that the radius is 10. Now suppose that the answers include both 10 and 20. The problem asked for the size in terms of the diameter, but you figured it out (quite reasonably) in terms of the radius. If, after all your work, you make the small slip of forgetting to convert, you get the problem wrong.

Now suppose you are asked for the percentage change if a price increases by 20% and then by 20% again. Suppose both 40 and 44 are among the answers. You reason that percents add and choose 40.

In both cases you have been distracted by a distractor, but the situations are very different. In the first case you were mostly right. In the second you were using a completely wrong approach.

In earlier years the AHSME had many distractors of both types. During 1983–88, due to strong feelings by many members of the AHSME Committee, most distractors of the first type (answers you might get if you did the problem mostly right and made just one small error) were eliminated. Distractors of the second type were not. Indeed they were encouraged – distractors were rarely chosen at random.

But what constitutes a small error? If you forget that $\sqrt{x}$ must be nonnegative, is that small, or have you neglected an important principle? These are matters the AHSME Committee argued about!

AHSME Scoring

The scoring change in 1986 (from "$30 + 4 * \text{Right} - \text{Wrong}$" to "$5 * \text{Right} + 2 * \text{Blank}$") was prompted primarily because of a concern about guessing. The data available to us indicated that on many questions even very good students were guessing wildly, i.e., randomly, or even badly, that is, with fewer right answers than if they had guessed at random. Indeed, as a *contest* strategy, random guessing was quite reasonable. Under the old scoring, random guessing would, on average, not change your score, and guessing after eliminating even one answer would, on average, increase your score. Furthermore, because the AHSME is hard, there was little embarrassment if guessing caused your score to go down, and much "fame" if by chance it made your score go way up. So why not go for broke?

Unfortunately, this contest strategy conflicts with the educational purpose, which for the student involves making one's strongest intellectual effort (including *educated* guessing) and for us involves accurate identification of the students with the greatest talent.

So we began by changing the instructions, urging students not to guess randomly. But this had little effect. The change in scoring was a more substantial change. First of all, it meant that random guessing would, on average, lower your score. Indeed, you have to eliminate three choices *before* a guess between the remaining two can be expected to raise your score (and the new instructions said this). Second, by awarding points for leaving a blank, instead of taking away points for a wrong answer, we hoped the new rules would interact with human psychology to discourage wild guessing.

There were some secondary arguments in favor of the change. For instance, even if students did not change their strategy, scores would rise. (That is, it is a little theorem that the same answers lead to an equal or higher score under the new rules). This allowed us to set the AIME cutoff at 100 (which sounds good, and matched our honor roll cutoff) instead of 95. Second, most scores would now be in the range 60–100, the typical range for scores on classroom tests. (Under much earlier scoring systems, it was not uncommon to get negative scores!)

There were some objections to the new rules. Some teachers felt we were rewarding passivity. (We felt we were rewarding good judgment about when to guess, and self-awareness about when you don't know what you are doing – how often do teachers see nonsense on student papers and wonder how the student failed to notice that it was nonsense?) But after the first year all went smoothly.

We don't claim this is the ideal scoring system for a multiple choice test. Indeed, there are some studies that suggest the best

system for such tests (i.e., the way to get the most accurate
measure of knowledge) is to allow students to select more than
one answer, say, by crossing out just those answers they are sure
are wrong. See "Scoring in Multiple-Choice Examinations, by G.
H. Pollard, *Math. Scientist* 10 (1985), pp. 93–97.

The AIME

The AIME was introduced for two reasons. First, by introducing
an exam intermediate in difficulty between AHSME and USAMO,
we could make the AHSME somewhat easier. Second, we hoped
the AIME would help us better identify the best students to
participate in the USAMO, which in turn is used to select the
American team for the International Mathematical Olympiad
(IMO), the most prestigious and difficult secondary mathematical
competition in the world. The USAMO is a 5-question, 3-hour
proof exam. Over the years there seemed to be only a weak
correlation between AHSME and USAMO scores. This is not
surprising. First, it is hard for any exam (in this case the
AHSME) to do fine discrimination over a narrow part of its range
(in this case the very top). Second, the formats of the two exams
are completely different. Speed and shrewd intuition pay off on
the AHSME (and they are valuable talents), but they are not
what the USAMO is about. Conversely, a student who is slower
but has a fine sense of logic could do well on the USAMO but
might never make it through the AHSME filter.

So the AMC members asked themselves: is it possible to devise
an exam that

1. is relatively easy to grade, so that a few thousand students
could take it but it could still be graded promptly; and

2. has problems more like Olympiad problems?

After much deliberation, we decided the answer was Yes. We
couldn't ask students to provide proofs and still grade a few
thousand papers. But we could make problems harder and give
much more time (twice as much time for half as many problems),
and we could eliminate guessing as a factor by requiring the exact
answer.

The answer – that was the most difficult issue! How could
we ask students to give the exact answer and still grade several
thousand papers promptly?

Our first idea was simply to leave a blank in which the student would write the answer. But this would not allow for machine grading. Worse, students might write correct answers in forms not easily recognized as correct. Suppose we were looking for the answer $\frac{1}{2}\sin^{-1}\frac{7}{9}$ and a student writes $\sin^{-1}\left(\frac{2\sqrt{2}}{3}\right) - \frac{\pi}{4}$? Is that right?

Our next idea was, for each problem, to state a required form for the answer. For instance, if our answer was $2 + \sqrt{3}$, then we might say: "Give integers a, b and c so that the answer is $a + b\sqrt{c}$. But we decided such instructions might give away too much information about the problem, and still would require hand grading.

So finally we decided that all answers would have to be integers from 0 to 999. Such answers could easily be encoded for machine scoring. Furthermore, guessing would be hard, even though this format was, strictly speaking, multiple choice.

The hard question, and the reason we demurred from this approach at first, was: could we still write a full range of questions given this constraint on the answers? It would be easy to write counting problems, and number theory problems, but what about algebra, geometry, trigonometry? Moreover, there was no point in allowing answers up to 999 if we were not going to use them. Would it be possible to make large, random-looking numbers show up as answers without making students do ugly, complicated arithmetic? Could we avoid having the mere fact that the answer is an integer be a clue?

You must judge whether we succeeded. The AIME Committee tried its hand at writing problems with the 0–999 constraint, and we felt we succeeded. As you work through the AIME problems, you will see some of the techniques we used to create integer answers.

AHSME-AIME Pairs

A good problem often suggests another similar one, sometimes harder, sometimes easier. This led us to wonder if we couldn't make pairs of problems that would appear, respectively on the AHSME and AIME, or even triples of problems to appear on AHSME, AIME and USAMO. If such linked problems appeared regularly, that would be an incentive, once the AHSME was over, for participants who were going on to the AIME to review the

AHSME very carefully and try to prepare for generalizations of AHSME problems.

It proved too difficult to produce such linked problems on a regular basis. Or rather, they were produced, but it was rare that both the AHSME Committee and the AIME Committee were equally enthusiastic about their halves of the pair. So no formal commitment was ever made to including pairs. Nonetheless, several did appear, as readers will see if they work through both sets of exams carefully.

Test Development

Readers might be curious to know how these tests are written, and older readers might be interested in getting involved.

The procedures settled upon during this period for developing the AHSME and AIME are much the same. About two years before an exam is to be given, the exam committee Chair calls for problems from the AMC panelists and exam committee members. (Panelists participate by mail only and are usually "trying out" for fuller involvement; committee members, say on the AHSME Subcommittee, also attend meetings where problems are fine-tuned and any policy issues concerning the exam, say the difficulty level or the scoring system, are discussed. Henceforth, panelists and subcommittee members will be called reviewers.) Each problem must be submitted in the appropriate format (e.g., multiple choice for the AHSME) along with a careful solution.

The Chair then makes a packet of all the contributions (sometimes after initial editing) and distributes it to the reviewers for evaluation. The Chair then collects the evaluations and uses them to produce a first draft. To make a good first draft often requires substantial editing and careful selection, including selection of a few extra problems in case some problems are later found to be unsuitable. In making the first draft, the Chair generally picks the best-liked problems, but he/she must also see to it that there is a progression from easy to very hard and that a number of different mathematical topics are covered. Sometimes the Chair must create another problem or two to fill gaps.

This first draft is sent to reviewers, who make written comments. On this basis, the Chair proposes various changes, and a subcommittee meeting takes place in which these and other changes are discussed. On this basis a second draft is developed and the process of review and meeting is repeated. Usually at this point the exam is ready to go to press, but occasionally a difficulty surfaces after this point and further changes have to be made. After a final proofreading review by one or two people associated with the AMC, but who have not previously been involved with producing this exam, the exam goes to press.

All along, a great deal of attention is paid to accuracy and clarity, since students taking these competitions do not get to ask questions about what the problems mean. As much attention is paid to the accuracy and clarity of the solutions as to the problems themselves.

Educators interested in participating in writing AMC examinations should contact the AMC Chair, currently Prof. Richard Gibbs, Fort Lewis College, Durango CO 81301. The AMC is especially interested in involving secondary-school teachers in this work. The best problems – fresh, intriguing, instructive, with multiple solution methods – are rare and cherished gems. Problem posing, like problem solving, is a special talent. But we are sure there are many wonderful problem posers out there, and the AMC would like to know you.

Thanks to All

We are the compilers of this book because the first of us was Chair of the AIME subcommittee during most of the period when these exams were produced and the second of us was the Chair of the AHSME subcommittee and of the AMC as a whole during that time. As explained earlier (test development), we had primary responsibility for putting the tests together, but we had lots of help. Only a small fraction of the problems and solutions in this book were created by one of us, and every problem and solution here benefited from the intense scrutiny of many people. So to all members of the AMC panel and committees during the time these exams were devised (approximately 60 people, some of whom have passed away) we give heartfelt thanks.

Editorial assistance, and encouragement, were provided by the Editorial Board of the New Mathematical Library series of the MAA, especially its current Chair, Underwood Dudley, its former Chairs, Paul Zorn and Ivan Niven, and its unnamed reviewers.

Other sorts of help were needed to put everything in book form. Here we would especially like to thank Mrs. Mary Lou McCullough, for her technical typing of much of the AIME materials and Professor Darrell J. Horwath of John Carroll University for his preparation of most of the figures accompanying the AIME problems and solutions. Once we finally got a manuscript to MAA headquarters, it was handled very promptly by the publication staff there, in particular, Donald Albers, Elaine Pedreira, and Beverly Ruedi.

Finally, special thanks go to Prof. Walter Mientka of the University of Nebraska at Lincoln, who was and is the Executive Director of the AMC. He is in charge of everything other than creating the exams (e.g., getting schools registered, getting exams printed and distributed, collecting score information, disseminating score information, publicizing the exams to the world, selling back material, managing finances...). Both of us found him immensely dedicated, and a pleasure to work with and know.

Closing Thoughts

It has been very gratifying to play a role in the AMC because of all the good it does to promote mathematics and mathematical talent. But our attachment goes deeper because of our personal histories.

The first of us says: Being a native of Hungary, I grew up on a rich diet of wonderful competitions and other problem-solving activities in the early 1950s. Therefore, following my emigration to the United States and the completion of my studies, it was natural for me to follow the tradition of Hungarian mathematicians and search for opportunities to make my own contributions to the competition scene. I was first invited to join the USAMO Subcommittee in 1976, became a member of the AHSME Panel in the same year, and assumed the chair of the AIME Subcommittee in 1982, following the untimely death of Professor John H. Staib of Drexel University.

The second of us says: Growing up in the United States, the AHSME was indeed the most special mathematical event of the year while I was in high school in the early 1960s. It was a joy, a challenge, and a focus of mathematical excitement. It surely had something to do with my becoming a mathematician. The AHSME continued to be all these things when I was a high school teacher in the early 70s and could offer it (and once the USAMO) to my good students. And so it was a great honor for me, and a great sense of coming full circle, when I was asked to join the Contest Committee in 1978.

In editing this volume, we are motivated by the hope that for some of you younger readers, these problems will be a similar source of joy and excitement. May you become mathematicians too, and take pleasure from problems like these all your life.

George Berzsenyi

Stephen Maurer

1983 AHSME

1. If $x \neq 0$, $\dfrac{x}{2} = y^2$ and $\dfrac{x}{4} = 4y$, then x equals

   (A) 8   (B) 16   (C) 32   (D) 64   (E) 128

2. Point P is outside circle C on the plane. At most how many points on C are 3 cm from P?

   (A) 1   (B) 2   (C) 3   (D) 4   (E) 8

3. Three primes, p, q and r, satisfy $p + q = r$ and $1 < p < q$. Then p equals

   (A) 2   (B) 3   (C) 7   (D) 13   (E) 17

4. In the adjoining plane figure, sides AF and CD are parallel, as are sides AB and FE, and sides BC and ED. Each side has length 1. Also, $\angle FAB = \angle BCD = 60°$. The area of the figure is

   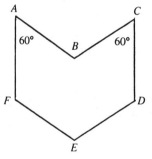

   (A) $\dfrac{\sqrt{3}}{2}$ (B) 1   (C) $\dfrac{3}{2}$

   (D) $\sqrt{3}$ (E) 2

5. Triangle ABC has a right angle at C. If $\sin A = 2/3$, then $\tan B$ is

   (A) $\dfrac{3}{5}$ (B) $\dfrac{\sqrt{5}}{3}$ (C) $\dfrac{2}{\sqrt{5}}$ (D) $\dfrac{\sqrt{5}}{2}$ (E) $\dfrac{5}{3}$

6. When

$$x^5, \quad x + \frac{1}{x} \quad \text{and} \quad 1 + \frac{2}{x} + \frac{3}{x^3}$$

   are multiplied, the product is a polynomial of degree

   (A) 2   (B) 3   (C) 6   (D) 7   (E) 8

1

7. Alice sells an item at \$10 less than the list price and receives 10% of her selling price as her commission. Bob sells the same item at \$20 less than the list price and receives 20% of his selling price as his commission. If they both get the same commission, then the list price is

(A) \$20   (B) \$30   (C) \$50   (D) \$70   (E) \$100

8. Let $f(x) = \dfrac{x+1}{x-1}$. Then for $x^2 \neq 1$, $f(-x)$ is

(A) $\dfrac{1}{f(x)}$ (B) $-f(x)$ (C) $\dfrac{1}{f(-x)}$ (D) $-f(-x)$ (E) $f(x)$

9. In a certain population the ratio of the number of women to the number of men is 11 to 10. If the average (arithmetic mean) age of the women is 34 and the average age of the men is 32, then the average age of the population is

(A) $32\frac{9}{10}$ (B) $32\frac{20}{21}$ (C) 33   (D) $33\frac{1}{21}$ (E) $33\frac{1}{10}$

10. Segment AB is both a diameter of a circle of radius 2 and a side of an equilateral triangle ABC. The circle also intersects AC and BC at points D and E, respectively. The length of AE is

(A) $\dfrac{3}{2}$ (B) $\dfrac{5}{3}$ (C) $\dfrac{\sqrt{3}}{2}$ (D) $\sqrt{3}$ (E) $\dfrac{2+\sqrt{3}}{2}$

11. Simplify $\sin(x-y)\cos y + \cos(x-y)\sin y$.

(A) 1   (B) $\sin x$ (C) $\cos x$ (D) $\sin x \cos 2y$

(E) $\cos x \cos 2y$

12. If $\log_7\big(\log_3(\log_2 x)\big) = 0$, then $x^{-1/2}$ equals

(A) $\dfrac{1}{3}$ (B) $\dfrac{1}{2\sqrt{3}}$ (C) $\dfrac{1}{3\sqrt{3}}$ (D) $\dfrac{1}{\sqrt{42}}$ (E) none of these

13. If $xy = a$, $xz = b$, and $yz = c$, and none of these quantities is 0, then $x^2 + y^2 + z^2$ equals

(A) $\dfrac{ab + ac + bc}{abc}$ (B) $\dfrac{a^2 + b^2 + c^2}{abc}$ (C) $\dfrac{(a + b + c)^2}{abc}$

(D) $\dfrac{(ab + ac + bc)^2}{abc}$ (E) $\dfrac{(ab)^2 + (ac)^2 + (bc)^2}{abc}$

14. The units digit of $3^{1001} 7^{1002} 13^{1003}$ is

(A) 1   (B) 3   (C) 5   (D) 7   (E) 9

15. Three balls marked 1, 2 and 3 are placed in an urn. One ball is drawn, its number is recorded, and then the ball is returned to the urn. This process is repeated and then repeated once more, and each ball is equally likely to be drawn on each occasion. If the sum of the numbers recorded is 6, what is the probability that the ball numbered 2 was drawn all three times?

(A) $\dfrac{1}{27}$ (B) $\dfrac{1}{8}$ (C) $\dfrac{1}{7}$ (D) $\dfrac{1}{6}$ (E) $\dfrac{1}{3}$

16. Let
$$x = .123456789101112\ldots998999,$$

where the digits are obtained by writing the integers 1 through 999 in order. The 1983rd digit to the right of the decimal point is

(A) 2   (B) 3   (C) 5   (D) 7   (E) 8

17. The diagram to the right shows several numbers in the complex plane. The circle is the unit circle centered at the origin. One of these numbers is the reciprocal of F. Which one?

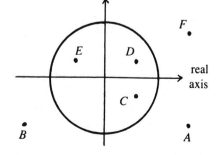

(A) A   (B) B   (C) C

(D) D   (E) E

18. Let f be a polynomial function such that, for all real x,

$$f(x^2+1) = x^4 + 5x^2 + 3.$$

For all real x, $f(x^2-1)$ is

(A) $x^4 + 5x^2 + 1$ (B) $x^4 + x^2 - 3$ (C) $x^4 - 5x^2 + 1$
(D) $x^4 + x^2 + 3$ (E) none of these

19. Point D is on side CB of triangle ABC. If

$$\angle CAD = \angle DAB = 60°, \quad AC = 3 \quad \text{and} \quad AB = 6,$$

then the length of AD is

(A) 2   (B) 2.5   (C) 3   (D) 3.5   (E) 4

20. If $\tan\alpha$ and $\tan\beta$ are the roots of $x^2 - px + q = 0$, and $\cot\alpha$ and $\cot\beta$ are the roots of $x^2 - rx + s = 0$, then rs is necessarily

(A) pq (B) $\dfrac{1}{pq}$ (C) $\dfrac{p}{q^2}$ (D) $\dfrac{q}{p^2}$ (E) $\dfrac{p}{q}$

21. Find the smallest positive number from the numbers below

(A) $10 - 3\sqrt{11}$ (B) $3\sqrt{11} - 10$ (C) $18 - 5\sqrt{13}$
(D) $51 - 10\sqrt{26}$ (E) $10\sqrt{26} - 51$

22. Consider the two functions

$$f(x) = x^2 + 2bx + 1 \quad \text{and} \quad g(x) = 2a(x+b),$$

where the variable x and the constants a and b are real numbers. Each such pair of constants a and b may be considered as a point (a, b) in an ab-plane. Let S be the set of such points (a, b) for which the graphs of $y = f(x)$ and $y = g(x)$ do **not** intersect (in the xy-plane). The area of S is

(A) 1   (B) π (C) 4   (D) 4π (E) infinite

23. In the adjoining figure the five circles are tangent to one another consecutively and to the lines L_1 and L_2. If the radius of the largest circle is 18 and that of the smallest one is 8, then the radius of the middle circle is

(A) 12   (B) 12.5

(C) 13   (D) 13.5

    (E) 14

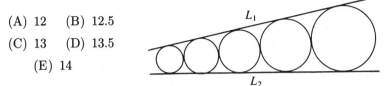

24. How many non-congruent right triangles are there such that the perimeter in cm and area in cm^2 are numerically equal?

(A) none   (B) 1   (C) 2   (D) 4   (E) infinitely many

25. If $60^a = 3$ and $60^b = 5$, then $12^{[(1-a-b)/2(1-b)]}$ is

(A) $\sqrt{3}$ (B) 2   (C) $\sqrt{5}$ (D) 3   (E) $\sqrt{12}$

26. The probability that event A occurs is $3/4$; the probability that event B occurs is $2/3$. Let p be the probability that both A and B occur. The smallest interval necessarily containing p is the interval

(A) $\left[\frac{1}{12}, \frac{1}{2}\right]$ (B) $\left[\frac{5}{12}, \frac{1}{2}\right]$ (C) $\left[\frac{1}{2}, \frac{2}{3}\right]$ (D) $\left[\frac{5}{12}, \frac{2}{3}\right]$

(E) $\left[\frac{1}{12}, \frac{2}{3}\right]$

27. A large sphere is on a horizontal field on a sunny day. At a certain time the shadow of the sphere reaches out a distance of 10 m from the point where the sphere touches the ground. At the same instant a meter stick (held vertically with one end on the ground) casts a shadow of length 2 m. What is the radius of the sphere in meters? (Assume the sun's rays are parallel and the meter stick is a line segment.)

(A) $\frac{5}{2}$ (B) $9 - 4\sqrt{5}$ (C) $8\sqrt{10} - 23$ (D) $6 - \sqrt{15}$

(E) $10\sqrt{5} - 20$

28. Triangle ABC in the figure has area 10. Points D, E and
F, all distinct from A, B and
C, are on sides AB, BC and
CA respectively, and $AD = 2$,
$DB = 3$. If triangle ABE and
quadrilateral $DBEF$ have equal
areas, then that area is

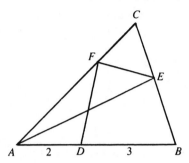

(A) 4   (B) 5   (C) 6

(D) $\frac{5}{3}\sqrt{10}$

(E) not uniquely determined

29. A point P lies in the same plane as a given square of side 1. Let
the vertices of the square, taken counterclockwise, be A, B, C and
D. Also, let the distances from P to A, B and C, respectively,
be u, v and w. What is the greatest distance that P can be from
D if $u^2 + v^2 = w^2$?

(A) $1 + \sqrt{2}$ (B) $2\sqrt{2}$ (C) $2 + \sqrt{2}$ (D) $3\sqrt{2}$ (E) $3 + \sqrt{2}$

30. Distinct points A and B are on a semicircle with diameter MN
and center C. The point P is on CN and

$\angle CAP = \angle CBP = 10°$.

If $\stackrel{\frown}{MA} = 40°$, then $\stackrel{\frown}{BN}$
equals

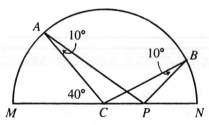

(A) 10°   (B) 15°

(C) 20°   (D) 25°

(E) 30°

1984 AHSME

1. $\dfrac{1000^2}{252^2 - 248^2}$ equals

    (A) 62,500   (B) 1000   (C) 500   (D) 250   (E) $\dfrac{1}{2}$

2. If x, y and $y - \dfrac{1}{x}$ are not 0, then

$$\frac{x - \dfrac{1}{y}}{y - \dfrac{1}{x}}$$

equals

    (A) 1   (B) $\dfrac{x}{y}$ (C) $\dfrac{y}{x}$ (D) $\dfrac{x}{y} - \dfrac{y}{x}$ (E) $xy - \dfrac{1}{xy}$

3. Let n be the smallest nonprime integer greater than 1 with no prime factor less than 10. Then

    (A) $100 < n \leq 110$ (B) $110 < n \leq 120$ (C) $120 < n \leq 130$
    (D) $130 < n \leq 140$ (E) $140 < n \leq 150$

4. A rectangle intersects a circle as shown: $AB = 4$, $BC = 5$ and
    $DE = 3$. Then EF equals

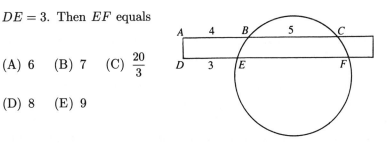

    (A) 6   (B) 7   (C) $\dfrac{20}{3}$

    (D) 8   (E) 9

5. The largest integer n for which $n^{200} < 5^{300}$ is

    (A) 8   (B) 9   (C) 10   (D) 11   (E) 12

6. In a certain school, there are three times as many boys as girls and nine times as many girls as teachers. Using the letters b, g, t to represent the number of boys, girls and teachers, respectively, then the total number of boys, girls and teachers can be represented by the expression

(A) $31b$ (B) $\frac{37}{27}b$ (C) $13g$ (D) $\frac{37}{27}g$ (E) $\frac{37}{27}t$

7. When Dave walks to school, he averages 90 steps per minute, each of his steps 75 cm long. It takes him 16 minutes to get to school His brother, Jack, going to the same school by the same route, averages 100 steps per minute, but his steps are only 60 cm long. How long does it take Jack to get to school?

(A) $14\frac{2}{9}$ min.   (B) 15 min.   (C) 18 min.   (D) 20 min.

(E) $22\frac{2}{9}$ min.

8. Figure $ABCD$ is a trapezoid with $AB \parallel DC$, $AB = 5$, $BC = 3\sqrt{2}$, $\angle BCD = 45°$ and $\angle CDA = 60°$.
The length of DC is

(A) $7 + \frac{2}{3}\sqrt{3}$ (B) 8

(C) $9\frac{1}{2}$ (D) $8 + \sqrt{3}$

(E) $8 + 3\sqrt{3}$

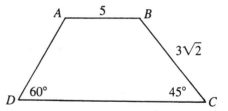

9. The number of digits in $4^{16}5^{25}$ (when written in the usual base 10 form) is

(A) 31   (B) 30   (C) 29   (D) 28   (E) 27

10. Four complex numbers lie at the vertices of a square in the complex plane. Three of the numbers are $1 + 2i$, $-2 + i$ and $-1 - 2i$. The fourth number is

(A) $2 + i$ (B) $2 - i$ (C) $1 - 2i$ (D) $-1 + 2i$ (E) $-2 - i$

11. A calculator has a key which replaces the displayed entry with its square, and another key which replaces the displayed entry with its reciprocal. Let y be the final result if one starts with an entry $x \neq 0$ and alternately squares and reciprocates n times each. Assuming the calculator is completely accurate (e.g., no roundoff or overflow), then y equals

(A) $x^{\left((-2)^n\right)}$ (B) x^{2n} (C) x^{-2n} (D) $x^{-(2^n)}$

(E) $x^{\left((-1)^n 2n\right)}$

12. If the sequence $\{a_n\}$ is defined by

$$a_1 = 2,$$
$$a_{n+1} = a_n + 2n \qquad (n \geq 1),$$

then a_{100} equals

(A) 9900   (B) 9902   (C) 9904   (D) 10100   (E) 10102

13. $\dfrac{2\sqrt{6}}{\sqrt{2}+\sqrt{3}+\sqrt{5}}$ equals

(A) $\sqrt{2}+\sqrt{3}-\sqrt{5}$ (B) $4-\sqrt{2}-\sqrt{3}$ (C) $\sqrt{2}+\sqrt{3}+\sqrt{6}-5$

(D) $\frac{1}{2}(\sqrt{2}+\sqrt{5}-\sqrt{3})$ (E) $\frac{1}{3}(\sqrt{3}+\sqrt{5}-\sqrt{2})$

14. The product of all real roots of the equation $x^{\log_{10} x} = 10$ is

(A) 1   (B) -1 (C) 10   (D) 10^{-1} (E) none of these

15. If $\sin 2x \sin 3x = \cos 2x \cos 3x$, then one value for x is

(A) $18°$ (B) $30°$ (C) $36°$ (D) $45°$ (E) $60°$

16. The function $f(x)$ satisfies $f(2+x) = f(2-x)$ for all real numbers x. If the equation $f(x) = 0$ has exactly four distinct real roots, then the sum of these roots is

(A) 0   (B) 2   (C) 4   (D) 6   (E) 8

17. A right triangle ABC with hypotenuse AB has side $AC = 15$. Altitude CH divides AB into segments AH and HB, with $HB = 16$. The area of $\triangle ABC$ is

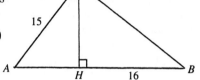

(A) 120   (B) 144   (C) 150

(D) 216   (E) $144\sqrt{5}$

18. A point (x, y) is to be chosen in the coordinate plane so that it is equally distant from the x-axis, the y-axis, and the line $x+y = 2$. Then x is

(A) $\sqrt{2} - 1$ (B) $\frac{1}{2}$ (C) $2 - \sqrt{2}$ (D) 1

(E) not uniquely determined

19. A box contains 11 balls, numbered $1, 2, 3, \ldots, 11$. If 6 balls are drawn simultaneously at random, what is the probability that the sum of the numbers on the balls drawn is odd?

(A) $\frac{100}{231}$ (B) $\frac{115}{231}$ (C) $\frac{1}{2}$ (D) $\frac{118}{231}$ (E) $\frac{6}{11}$

20. The number of distinct solutions of the equation $\big| x - |2x+1| \big| = 3$ is

(A) 0   (B) 1   (C) 2   (D) 3   (E) 4

21. The number of triples (a, b, c) of positive integers which satisfy the simultaneous equations

$$ab + bc = 44,$$
$$ac + bc = 23,$$

is

(A) 0   (B) 1   (C) 2   (D) 3   (E) 4

22. Let a and c be fixed positive numbers. For each real number t let (x_t, y_t) be the vertex of the parabola $y = ax^2 + tx + c$. If the set of vertices (x_t, y_t) for all real values of t is graphed in the plane, the graph is

(A) a straight line   (B) a parabola

(C) part, but not all, of a parabola

(D) one branch of a hyperbola   (E) none of these

23. $\dfrac{\sin 10° + \sin 20°}{\cos 10° + \cos 20°}$ equals

(A) $\tan 10° + \tan 20°$ (B) $\tan 30°$ (C) $\dfrac{1}{2}(\tan 10° + \tan 20°)$

(D) $\tan 15°$ (E) $\dfrac{1}{4}\tan 60°$

24. If a and b are positive real numbers and each of the equations

$$x^2 + ax + 2b = 0 \quad \text{and} \quad x^2 + 2bx + a = 0$$

has real roots, then the smallest possible value of $a + b$ is

(A) 2   (B) 3   (C) 4   (D) 5   (E) 6

25. The total area of all the faces of a rectangular solid is $22\,\text{cm}^2$, and the total length of all its edges is $24\,\text{cm}$. Then the length in cm of any one of its internal diagonals is

(A) $\sqrt{11}$ (B) $\sqrt{12}$ (C) $\sqrt{13}$ (D) $\sqrt{14}$

(E) not uniquely determined

26. In the obtuse triangle ABC, $AM = MB$, $MD \perp BC$, $EC \perp BC$. If the area of $\triangle ABC$ is 24, then the area of $\triangle BED$ is

(A) 9   (B) 12   (C) 15   (D) 18

(E) not uniquely determined

27. In $\triangle ABC$, D is on AC and F is on BC. Also, $AB \perp AC$, $AF \perp BC$, and $BD = DC = FC = 1$. Find AC

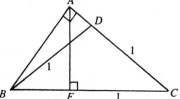

   (A) $\sqrt{2}$ (B) $\sqrt{3}$ (C) $\sqrt[3]{2}$

   (D) $\sqrt[3]{3}$ (E) $\sqrt[4]{3}$

28. The number of distinct pairs of integers (x, y) such that

$$0 < x < y \quad \text{and} \quad \sqrt{1984} = \sqrt{x} + \sqrt{y}$$

is

   (A) 0   (B) 1   (C) 3   (D) 4   (E) 7

29. Find the largest value of $\dfrac{y}{x}$ for pairs of real numbers (x, y) which satisfy

$$(x - 3)^2 + (y - 3)^2 = 6.$$

   (A) $3 + 2\sqrt{2}$ (B) $2 + \sqrt{3}$ (C) $3\sqrt{3}$ (D) 6   (E) $6 + 2\sqrt{3}$

30. For any complex number $w = a + bi$, $|w|$ is defined to be the real number $\sqrt{a^2 + b^2}$. If $w = \cos 40° + i \sin 40°$, then

$$|w + 2w^2 + 3w^3 + \cdots + 9w^9|^{-1}$$

equals

   (A) $\dfrac{1}{9} \sin 40°$ (B) $\dfrac{2}{9} \sin 20°$ (C) $\dfrac{1}{9} \cos 40°$ (D) $\dfrac{1}{18} \cos 20°$

   (E) none of these

1985 AHSME

1. If $2x + 1 = 8$, then $4x + 1 =$

   (A) 15   (B) 16   (C) 17   (D) 18   (E) 19

2. In an arcade game, the "monster" is the shaded sector of a circle of radius 1 cm, as shown in the figure. The missing piece (the mouth) has central angle 60°. What is the perimeter of the monster in cm?

   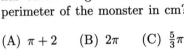

   (A) $\pi + 2$ (B) 2π (C) $\frac{5}{3}\pi$

   (D) $\frac{5}{6}\pi + 2$ (E) $\frac{5}{3}\pi + 2$

3. In right $\triangle ABC$ with legs 5 and 12, arcs of circles are drawn, one with center A and radius 12, the other with center B and radius 5. They intersect the hypotenuse in M and N. Then MN has length

   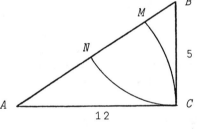

   (A) 2   (B) $\frac{13}{5}$ (C) 3

   (D) 4   (E) $\frac{24}{5}$

4. A large bag of coins contains pennies, dimes and quarters. There are twice as many dimes as pennies and three times as many quarters as dimes. An amount of money which could be in the bag is

   (A) \$306   (B) \$333   (C) \$342   (D) \$348   (E) \$360

5. Which terms must be removed from the sum

   $$\frac{1}{2} + \frac{1}{4} + \frac{1}{6} + \frac{1}{8} + \frac{1}{10} + \frac{1}{12}$$

   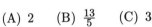

   if the sum of the remaining terms is to equal 1?

   (A) $\frac{1}{4}$ and $\frac{1}{8}$ (B) $\frac{1}{4}$ and $\frac{1}{12}$ (C) $\frac{1}{8}$ and $\frac{1}{12}$ (D) $\frac{1}{6}$ and $\frac{1}{10}$

   (E) $\frac{1}{8}$ and $\frac{1}{10}$

6. One student in a class of boys and girls is to be chosen to represent the class. Each student is equally likely to be chosen and the probability that a boy is chosen is 2/3 of the probability that a girl is chosen. The ratio of the number of boys to the total number of boys and girls is

(A) 1/3   (B) 2/5   (C) 1/2   (D) 3/5   (E) 2/3

7. In some computer languages (such as APL), when there are no parentheses in an algebraic expression, the operations are grouped from right to left. Thus, $a \times b - c$ in such languages means the same as $a(b-c)$ in ordinary algebraic notation. If $a \div b - c + d$ is evaluated in such a language, the result in ordinary algebraic notation would be

(A) $\dfrac{a}{b-c+d}$ (B) $\dfrac{a}{b-c-d}$ (C) $\dfrac{d+c-b}{a}$ (D) $\dfrac{a}{b-c+d}$

(E) $\dfrac{a}{b-c-d}$

8. Let a, a', b, b' be real numbers with a and a' nonzero. The solution to $ax + b = 0$ is less than the solution to $a'x + b' = 0$ if and only if

(A) $a'b < ab'$ (B) $ab' < a'b$ (C) $ab < a'b'$ (D) $\dfrac{b}{a} < \dfrac{b'}{a'}$

(E) $\dfrac{b}{a'} < \dfrac{b}{a}$

9. The odd positive integers, $1, 3, 5, 7, \ldots$, are arranged in five columns continuing with the pattern shown on the right. Counting from the left, the column in which 1985 appears is the

(A) first   (B) second

(C) third   (D) fourth

(E) fifth

	1	3	5	7
15	13	11	9	
	17	19	21	23
31	29	27	25	
	33	35	37	39
47	45	43	41	
	49	51	53	55
·	·	·	·	
	·	·	·	·
·	·	·	·	

10. An arbitrary circle can intersect the graph of $y = \sin x$ in

(A) at most 2 points   (B) at most 4 points

(C) at most 6 points   (D) at most 8 points

(E) more than 16 points

11. How many distinguishable rearrangements of the letters in CONTEST have both the vowels first? (For instance, OETCNST is one such arrangement, but OTETSNC is not.)

(A) 60   (B) 120   (C) 240   (D) 720   (E) 2520

12. Let p, q and r be distinct prime numbers, where 1 is not considered a prime. Which of the following is the smallest positive perfect cube having $n = pq^2r^4$ as a divisor?

(A) $p^8q^8r^8$ (B) $(pq^2r^2)^3$ (C) $(p^2q^2r^2)^3$ (D) $(pqr^2)^3$

(E) $4p^3q^3r^3$

13. Pegs are put in a board 1 unit apart both horizontally and vertically. A rubber band is stretched over 4 pegs as shown in the figure, forming a quadrilateral. Its area in square units is

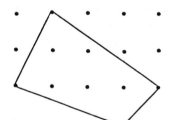

(A) 4   (B) 4.5   (C) 5

(D) 5.5   (E) 6

14. Exactly three of the interior angles of a convex polygon are obtuse. What is the maximum number of sides of such a polygon?

(A) 4   (B) 5   (C) 6   (D) 7   (E) 8

15. If a and b are positive numbers such that $a^b = b^a$ and $b = 9a$, then the value of a is

(A) 9   (B) 1/9   (C) $\sqrt[9]{9}$ (D) $\sqrt[3]{9}$ (E) $\sqrt[4]{3}$

16. If $A = 20°$ and $B = 25°$, then the value of $(1 + \tan A)(1 + \tan B)$ is

(A) $\sqrt{3}$ (B) 2   (C) $1 + \sqrt{2}$ (D) $2(\tan A + \tan B)$

(E) none of these

17. Diagonal DB of rectangle $ABCD$ is divided into three segments of length 1 by parallel lines L and L' that pass through A and C and are perpendicular to DB. The area of $ABCD$, rounded to one decimal place, is

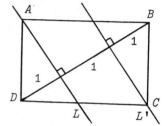

(A) 4.1   (B) 4.2   (C) 4.3

(D) 4.4   (E) 4.5

18. Six bags of marbles contain 18, 19, 21, 23, 25 and 34 marbles, respectively. One bag contains chipped marbles only. The other 5 bags contain no chipped marbles. Jane takes three of the bags and George takes two of the others. Only the bag of chipped marbles remains. If Jane gets twice as many marbles as George, how many chipped marbles are there?

(A) 18   (B) 19   (C) 21   (D) 23   (E) 25

19. Consider the graphs of $y = Ax^2$ and $y^2 + 3 = x^2 + 4y$, where A is a *positive* constant and x and y are real variables. In how many points do the two graphs intersect?

(A) exactly 4   (B) exactly 2

(C) at least 1, but the number varies for different positive

values of A

(D) 0 for at least one positive value of A

(E) none of these

20. A wooden cube with edge length n units (where n is an integer > 2) is painted black all over. By slices parallel to its faces, the cube is cut into n^3 smaller cubes each of unit edge length. If the number of smaller cubes with just one face painted black is equal to the number of smaller cubes completely free of paint, what is n?

(A) 5   (B) 6   (C) 7   (D) 8   (E) none of these

21. How many integers x satisfy the equation

$$(x^2 - x - 1)^{x+2} = 1 ?$$

(A) 2   (B) 3   (C) 4   (D) 5   (E) none of these

22. In a circle with center O, AD is a diameter, ABC is a chord, $BO = 5$ and $\angle ABO = \overset{\frown}{CD} = 60°$. Then the length of BC is

(A) 3   (B) $3 + \sqrt{3}$

(C) $5 - \frac{\sqrt{3}}{2}$ (D) 5

(E) none of the above

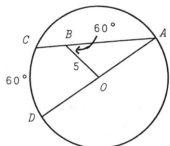

23. If

$$x = \frac{-1 + i\sqrt{3}}{2} \quad \text{and} \quad y = \frac{-1 - i\sqrt{3}}{2},$$

where $i^2 = -1$, then which of the following is **not** correct?

(A) $x^5 + y^5 = -1$ (B) $x^7 + y^7 = -1$ (C) $x^9 + y^9 = -1$
(D) $x^{11} + y^{11} = -1$ (E) $x^{13} + y^{13} = -1$

24. A non-zero digit is chosen in such a way that the probability of choosing digit d is $\log_{10}(d+1) - \log_{10} d$. The probability that the digit 2 is chosen is exactly 1/2 the probability that the digit chosen is in the set

(A) $\{2, 3\}$ (B) $\{3, 4\}$ (C) $\{4, 5, 6, 7, 8\}$ (D) $\{5, 6, 7, 8, 9\}$
(E) $\{4, 5, 6, 7, 8, 9\}$

25. The volume of a certain rectangular solid is $8\,\text{cm}^3$, its total surface area is $32\,\text{cm}^2$, and its three dimensions are in geometric progression. The sum of the lengths in cm of all the edges of this solid is

(A) 28   (B) 32   (C) 36   (D) 40   (E) 44

26. Find the least positive integer n for which $\dfrac{n-13}{5n+6}$ is a non-zero reducible fraction.

(A) 45   (B) 68   (C) 155   (D) 226   (E) none of these

27. Consider a sequence $x_1, x_2, x_3, \ldots$, defined by:
$$x_1 = \sqrt[3]{3},$$
$$x_2 = \left(\sqrt[3]{3}\right)^{\sqrt[3]{3}},$$
and in general
$$x_n = (x_{n-1})^{\sqrt[3]{3}} \quad \text{for} \quad n > 1.$$
What is the smallest value of n for which x_n is an integer?

(A) 2   (B) 3   (C) 4   (D) 9   (E) 27

28. In $\triangle ABC$, we have $\angle C = 3\angle A$, $a = 27$ and $c = 48$. What is b?

(A) 33   (B) 35

(C) 37   (D) 39

(E) not uniquely determined

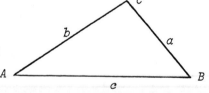

29. In their base 10 representations, the integer a consists of a sequence of 1985 eights and the integer b consists of a sequence of 1985 fives. What is the sum of the digits of the base 10 representation of the integer $9ab$?

(A) 15880   (B) 17856   (C) 17865   (D) 17874   (E) 19851

30. Let $\lfloor x \rfloor$ be the greatest integer less than or equal to x. Then the number of real solutions to $4x^2 - 40\lfloor x \rfloor + 51 = 0$ is

(A) 0   (B) 1   (C) 2   (D) 3   (E) 4

1986 AHSME

1. $[x - (y-z)] - [(x-y) - z] =$

   (A) $2y$ (B) $2z$ (C) $-2y$ (D) $-2z$ (E) 0

2. If the line L in the xy-plane has half the slope and twice the y-intercept of the line $y = \frac{2}{3}x + 4$, then an equation for L is

   (A) $y = \frac{1}{3}x + 8$ (B) $y = \frac{4}{3}x + 2$ (C) $y = \frac{1}{3}x + 4$

   (D) $y = \frac{4}{3}x + 4$ (E) $y = \frac{1}{3}x + 2$

3. In the figure, $\triangle ABC$ has a right angle at C and $\angle A = 20°$. If BD is the bisector of $\angle ABC$, then $\angle BDC =$

   (A) $40°$ (B) $45°$ (C) $50°$

   (D) $55°$ (E) $60°$

   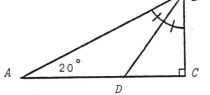

4. Let S be the statement

   "If the sum of the digits of the whole number n is divisible by 6, then n is divisible by 6."

   A value of n which shows S to be false is

   (A) 30   (B) 33   (C) 40   (D) 42   (E) none of these

5. Simplify $\left(\sqrt[6]{27} - \sqrt{6\frac{3}{4}} \right)^2$.

   (A) $\frac{3}{4}$ (B) $\frac{\sqrt{3}}{2}$ (C) $\frac{3\sqrt{3}}{4}$ (D) $\frac{3}{2}$ (E) $\frac{3\sqrt{3}}{2}$

19

6. Using a table of a certain height, two identical blocks of wood are placed as shown in Figure 1. Length r is found to be 32 inches. After rearranging the blocks as in Figure 2, length s is found to be 28 inches. How high is the table?

(A) 28 inches

(B) 29 inches

(C) 30 inches

(D) 31 inches

(E) 32 inches

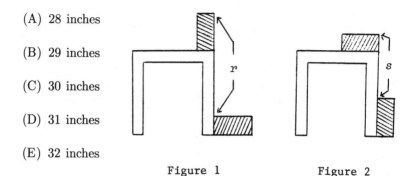

Figure 1   Figure 2

7. The sum of the greatest integer less than or equal to x and the least integer greater than or equal to x is 5. The solution set for x is

(A) $5/2$ (B) $\{x \mid 2 \leq x \leq 3\}$ (C) $\{x \mid 2 \leq x < 3\}$

(D) $\{x \mid 2 < x \leq 3\}$ (E) $\{x \mid 2 < x < 3\}$

8. The population of the United States in 1980 was 226,504,825. The area of the country is 3,615,122 square miles. There are $(5280)^2$ square feet in one square mile. Which number below best approximates the average number of square feet per person?

(A) 5,000   (B) 10,000   (C) 50,000   (D) 100,000

(E) 500,000

9. The product

$$\left(1 - \frac{1}{2^2}\right)\left(1 - \frac{1}{3^2}\right) \cdots \left(1 - \frac{1}{9^2}\right)\left(1 - \frac{1}{10^2}\right)$$

equals

(A) $\dfrac{5}{12}$ (B) $\dfrac{1}{2}$ (C) $\dfrac{11}{20}$ (D) $\dfrac{2}{3}$ (E) $\dfrac{7}{10}$

10. The 120 permutations of AHSME are arranged in dictionary order, as if each were an ordinary five-letter word. The last letter of the 86th word in this list is

(A) A   (B) H   (C) S   (D) M   (E) E

11. In $\triangle ABC$, $AB = 13$, $BC = 14$ and $CA = 15$. Also, M is the midpoint of side AB and H is the foot of the altitude from A to BC. The length of HM is

(A) 6   (B) 6.5   (C) 7

(D) 7.5   (E) 8

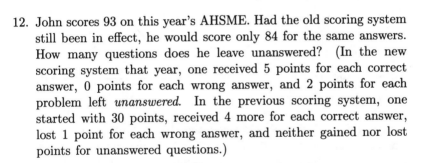

12. John scores 93 on this year's AHSME. Had the old scoring system still been in effect, he would score only 84 for the same answers. How many questions does he leave unanswered? (In the new scoring system that year, one received 5 points for each correct answer, 0 points for each wrong answer, and 2 points for each problem left *unanswered*. In the previous scoring system, one started with 30 points, received 4 more for each correct answer, lost 1 point for each wrong answer, and neither gained nor lost points for unanswered questions.)

(A) 6   (B) 9   (C) 11   (D) 14   (E) not uniquely determined

13. A parabola $y = ax^2 + bx + c$ has vertex (4,2). If (2,0) is on the parabola, then abc equals

(A) -12 (B) -6 (C) 0   (D) 6   (E) 12

14. Suppose hops, skips and jumps are specific units of length. If b hops equals c skips, d jumps equals e hops, and f jumps equals g meters, then one meter equals how many skips?

(A) $\dfrac{bdg}{cef}$ (B) $\dfrac{cdf}{beg}$ (C) $\dfrac{cdg}{bef}$ (D) $\dfrac{cef}{bdg}$ (E) $\dfrac{ceg}{bdf}$

15. A student attempted to compute the average, A, of x, y and z by computing the average of x and y, and then computing the average of the result and z. Whenever $x < y < z$, the student's final result is

(A) correct

(B) always less than A

(C) always greater than A

(D) sometimes less than A and sometimes equal to A

(E) sometimes greater than A and sometimes equal to A

16. In $\triangle ABC$, $AB = 8$, $BC = 7$, $CA = 6$ and side BC is extended, as shown in the figure, to a point P so that $\triangle PAB$ is similar to $\triangle PCA$. The length of PC is

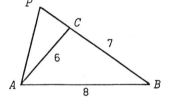

(A) 7   (B) 8   (C) 9

(D) 10   (E) 11

17. A drawer in a darkened room contains 100 red socks, 80 green socks, 60 blue socks and 40 black socks. A youngster selects socks one at a time from the drawer but is unable to see the color of the socks drawn. What is the smallest number of socks that must be selected to guarantee that the selection contains at least 10 pairs? (A pair of socks is two socks of the same color. No sock may be counted in more than one pair.)

(A) 21   (B) 23   (C) 24   (D) 30   (E) 50

18. A plane intersects a right circular cylinder of radius 1 forming an ellipse. If the major axis of the ellipse is 50% longer than the minor axis, the length of the major axis is

(A) 1   (B) $\dfrac{3}{2}$ (C) 2   (D) $\dfrac{9}{4}$ (E) 3

19. A park is in the shape of a regular hexagon 2 km on a side. Starting at a corner, Alice walks along the perimeter of the park for a distance of 5 km. How many kilometers is she from her starting point?

(A) $\sqrt{13}$ (B) $\sqrt{14}$ (C) $\sqrt{15}$ (D) $\sqrt{16}$ (E) $\sqrt{17}$

20. Suppose x and y are inversely proportional and positive. If x increases by $p\%$, then y decreases by

(A) $p\%$ (B) $\dfrac{p}{1+p}\%$ (C) $\dfrac{100}{p}\%$ (D) $\dfrac{p}{100+p}\%$

(E) $\dfrac{100p}{100+p}\%$

21. In the configuration below, θ is measured in radians, C is the center of the circle, BCD and ACE are line segments, and AB is tangent to the circle at A.

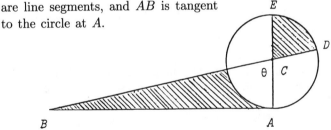

A necessary and sufficient condition for the equality of the two shaded areas, given $0 < \theta < \pi/2$, is

(A) $\tan\theta = \theta$ (B) $\tan\theta = 2\theta$ (C) $\tan\theta = 4\theta$

(D) $\tan 2\theta = \theta$ (E) $\tan\dfrac{\theta}{2} = \theta$

22. Six distinct integers are picked at random from $\{1, 2, 3, \ldots, 10\}$. What is the probability that, among those selected, the second smallest is 3?

(A) 1/60   (B) 1/6   (C) 1/3   (D) 1/2   (E) none of these

23. Let
$$N = 69^5 + 5 \cdot 69^4 + 10 \cdot 69^3 + 10 \cdot 69^2 + 5 \cdot 69 + 1.$$
How many positive integers are factors of N?

(A) 3   (B) 5   (C) 69   (D) 125   (E) 216

24. Let $p(x) = x^2 + bx + c,$ where b and c are integers. If $p(x)$ is a factor of both
$$x^4 + 6x^2 + 25 \quad \text{and} \quad 3x^4 + 4x^2 + 28x + 5,$$
what is $p(1)$?

(A) 0   (B) 1   (C) 2   (D) 4   (E) 8

25. If $\lfloor x \rfloor$ is the greatest integer less than or equal to x, then
$$\sum_{N=1}^{1024} \lfloor \log_2 N \rfloor =$$

(A) 8192   (B) 8204   (C) 9218   (D) $\lfloor \log_2(1024!) \rfloor$
(E) none of these

26. It is desired to construct a right triangle in the coordinate plane so that its legs are parallel to the x and y axes and so that the medians to the midpoints of the legs lie on the lines $y = 3x + 1$ and $y = mx + 2$. The number of different constants m for which such a triangle exists is

(A) 0   (B) 1   (C) 2   (D) 3   (E) more than 3

27. In the adjoining figure, AB is a diameter of the circle, CD is a chord parallel to AB, and AC intersects BD at E, with $\angle AED = \alpha$. The ratio of the area of $\triangle CDE$ to that of $\triangle ABE$ is

(A) $\cos\alpha$ (B) $\sin\alpha$

(C) $\cos^2\alpha$ (D) $\sin^2\alpha$

(E) $1 - \sin\alpha$

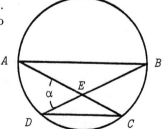

28. $ABCDE$ is a regular pentagon. AP, AQ and AR are the perpendiculars dropped from A onto CD, CB extended and DE extended, respectively. Let O be the center of the pentagon. If $OP = 1$, then $AO + AQ + AR$ equals

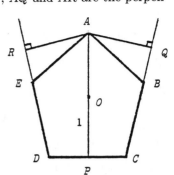

(A) 3 (B) $1 + \sqrt{5}$ (C) 4

(D) $2 + \sqrt{5}$ (E) 5

29. Two of the altitudes of the scalene triangle ABC have length 4 and 12. If the length of the third altitude is also an integer, what is the biggest it can be?

(A) 4 (B) 5 (C) 6 (D) 7 (E) none of these

30. The number of real solutions (x, y, z, w) of the simultaneous equations

$$2y = x + \frac{17}{x}, \quad 2z = y + \frac{17}{y}, \quad 2w = z + \frac{17}{z}, \quad 2x = w + \frac{17}{w}$$

is

(A) 1 (B) 2 (C) 4 (D) 8 (E) 16

1987 AHSME

1. $(1 + x^2)(1 - x^3)$ equals

   (A) $1 - x^5$ (B) $1 - x^6$ (C) $1 + x^2 - x^3$
   (D) $1 + x^2 - x^3 - x^5$ (E) $1 + x^2 - x^3 - x^6$

2. As shown in the figure, a triangular corner with side lengths
   $DB = EB = 1$ is cut from equilateral
   triangle ABC of side length 3. The
   perimeter of the remaining quadrilateral $ADEC$ is

   (A) 6   (B) $6\frac{1}{2}$ (C) 7

   (D) $7\frac{1}{2}$ (E) 8

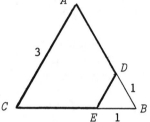

3. How many primes less than 100 have 7 as the ones digit? (Assume
   the usual base 10 representation.)

   (A) 4   (B) 5   (C) 6   (D) 7   (E) 8

4. $\dfrac{2^1 + 2^0 + 2^{-1}}{2^{-2} + 2^{-3} + 2^{-4}}$ equals

   (A) 6   (B) 8   (C) $\dfrac{31}{2}$ (D) 24   (E) 512

5. A student recorded the exact percentage frequency distribution
   for a set of measurements, as
   shown to the right. However,
   the student neglected to indicate
   N, the total number of measurements. What is the smallest possible value of N?

   (A) 5   (B) 8   (C) 16

   (D) 25   (E) 50

measured value	percent frequency
0	12.5
1	0
2	50
3	25
4	12.5
	100

6. In the $\triangle ABC$ shown, D is some interior point, and x, y, z, w are the measures of angles in degrees. Solve for x in terms of y, z and w.

(A) $w - y - z$

(B) $w - 2y - 2z$

(C) $180 - w - y - z$

(D) $2w - y - z$

(E) $180 - w + y + z$

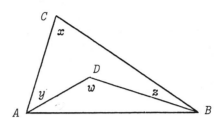

7. If $a - 1 = b + 2 = c - 3 = d + 4$, which of the four quantities a, b, c, d is the largest?

(A) a (B) b (C) c (D) d (E) no one is always largest

8. In the figure the sum of the distances AD and BD is

(A) between 10 and 11

(B) 12

(C) between 15 and 16

(D) between 16 and 17

(E) 17

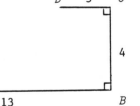

9. The first four terms of an arithmetic sequence are a, x, b, $2x$. The ratio of a to b is

(A) 1/4 (B) 1/3 (C) 1/2 (D) 2/3 (E) 2

10. How many ordered triples (a, b, c) of non-zero real numbers have the property that each number is the product of the other two?

(A) 1 (B) 2 (C) 3 (D) 4 (E) 5

11. Let c be a constant. The simultaneous equations

$$x - y = 2,$$
$$cx + y = 3$$

have a solution (x, y) inside Quadrant I if and only if

(A) $c = -1$ (B) $c > -1$ (C) $c < 3/2$ (D) $0 < c < 3/2$

(E) $-1 < c < 3/2$

12. In an office, at various times during the day the boss gives the secretary a letter to type, each time putting the letter on top of the pile in the secretary's in-box. When there is time, the secretary takes the top letter off the pile and types it. If there are five letters in all, and the boss delivers them in the order 1 2 3 4 5, which of the following could *not* be the order in which the secretary types them?

(A) 1 2 3 4 5   (B) 2 4 3 5 1   (C) 3 2 4 1 5

(D) 4 5 2 3 1   (E) 5 4 3 2 1

13. A long piece of paper 5 cm wide is made into a roll for cash registers by wrapping it 600 times around a cardboard tube of diameter 2 cm, forming a roll 10 cm in diameter. Approximate the length of the paper in meters. (Pretend the paper forms 600 concentric circles with diameters evenly spaced from 2 cm to 10 cm.)

(A) 36π (B) 45π (C) 60π (D) 72π (E) 90π

14. $ABCD$ is a square and M and N are the midpoints of BC and CD respectively. Then $\sin\theta =$

(A) $\dfrac{\sqrt{5}}{5}$ (B) $\dfrac{3}{5}$ (C) $\dfrac{\sqrt{10}}{5}$

(D) $\dfrac{4}{5}$ (E) none of these

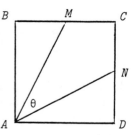

15. If (x, y) is a solution to the system

$$xy = 6 \quad \text{and} \quad x^2 y + xy^2 + x + y = 63,$$

find $x^2 + y^2$.

(A) 13   (B) $\dfrac{1173}{32}$ (C) 55   (D) 69   (E) 81

16. A cryptographer devises the following method for encoding positive integers. First, the integer is expressed in base 5. Second, a 1-to-1 correspondence is established between the digits that appear in the expressions in base 5 and the elements of the set $\{V, W, X, Y, Z\}$. Using this correspondence, the cryptographer finds that three consecutive integers in increasing order are coded as VYZ, VYX, VVW, respectively. What is the base-10 expression for the integer coded as XYZ?

(A) 48   (B) 71   (C) 82   (D) 108   (E) 113

17. In a mathematics competition, the sum of the scores of Bill and Dick equalled the sum of the scores of Ann and Carol. If the scores of Bill and Carol had been interchanged, then the sum of the scores of Ann and Carol would have exceeded the sum of the scores of the other two. Also, Dick's score exceeded the sum of the scores of Bill and Carol. Determine the order in which the four contestants finished, from highest to lowest. Assume all scores were nonnegative.

(A) Dick, Ann, Carol, Bill   (B) Dick, Ann, Bill, Carol

(C) Dick, Carol, Bill, Ann   (D) Ann, Dick, Carol, Bill

(E) Ann, Dick, Bill, Carol

18. It takes A algebra books (all the same thickness) and H geometry books (all the same thickness, which is greater than that of an algebra book) to completely fill a certain shelf. Also, S of the algebra books and M of the geometry books would fill the same shelf. Finally, E of the algebra books alone would fill this shelf. Given that A, H, S, M, E are distinct positive integers, it follows that E is

(A) $\dfrac{AM + SH}{M + H}$ (B) $\dfrac{AM^2 + SH^2}{M^2 + H^2}$ (C) $\dfrac{AH - SM}{M - H}$

(D) $\dfrac{AM - SH}{M - H}$ (E) $\dfrac{AM^2 - SH^2}{M^2 - H^2}$

19. Which of the following is closest to $\sqrt{65} - \sqrt{63}$?

(A) .12   (B) .13   (C) .14   (D) .15   (E) .16

20. Evaluate

$$\log_{10}(\tan 1°) + \log_{10}(\tan 2°) + \log_{10}(\tan 3°) + \cdots +$$
$$\log_{10}(\tan 88°) + \log_{10}(\tan 89°).$$

(A) 0   (B) $\frac{1}{2}\log_{10}(\frac{1}{2}\sqrt{3})$ (C) $\frac{1}{2}\log_{10} 2$ (D) 1
(E) none of these

21. There are two natural ways to inscribe a square in a given isosceles right triangle. If it is done as in Figure 1 below, then one finds that the area of the square is $441\,\text{cm}^2$. What is the area (in cm^2) of the square inscribed in the same $\triangle ABC$ as shown in Figure 2 below?

(A) 378

(B) 392

(C) 400

(D) 441

(E) 484

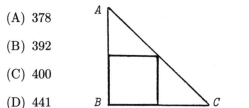

Figure 1   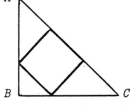 Figure 2

22. A ball was floating in a lake when the lake froze. The ball was removed (without breaking the ice), leaving a hole 24 cm across at the top and 8 cm deep. What was the radius of the ball (in centimeters)?

(A) 8   (B) 12   (C) 13   (D) $8\sqrt{3}$ (E) $6\sqrt{6}$

23. If p is a prime and both roots of $x^2 + px - 444p = 0$ are integers, then

(A) $1 < p \le 11$ (B) $11 < p \le 21$ (C) $21 < p \le 31$
(D) $31 < p \le 41$ (E) $41 < p \le 51$

24. How many polynomial functions f of degree ≥ 1 satisfy

$$f(x^2) = [f(x)]^2 = f(f(x)) \, ?$$

(A) 0   (B) 1   (C) 2   (D) finitely many but more than 2
(E) infinitely many

25. ABC is a triangle: $A = (0,0)$, $B = (36,15)$ and both the coordinates of C are integers. What is the minimum area $\triangle ABC$ can have?

(A) 1/2   (B) 1   (C) 3/2   (D) 13/2
(E) there is no minimum

26. The amount 2.5 is split into two nonnegative real numbers uniformly at random, for instance, into 2.143 and .357, or into $\sqrt{3}$ and $2.5 - \sqrt{3}$. Then each number is rounded to its nearest integer, for instance, 2 and 0 in the first case above, 2 and 1 in the second. What is the probability that the two integers sum to 3?

(A) 1/4   (B) 2/5   (C) 1/2   (D) 3/5   (E) 3/4

27. A cube of cheese $C = \{ (x,y,z) \mid 0 \leq x,y,z \leq 1 \}$ is cut along the planes $x = y$, $y = z$ and $z = x$. How many pieces are there? (No cheese is moved until all three cuts are made.)

(A) 5   (B) 6   (C) 7   (D) 8   (E) 9

28. Let a, b, c, d be real numbers. Suppose that all the roots of $z^4 + az^3 + bz^2 + cz + d = 0$ are complex numbers lying on a circle in the complex plane centered at $0 + 0i$ and having radius 1. The sum of the reciprocals of the roots is necessarily

(A) a (B) b (C) c (D) $-a$ (E) $-b$

29. Consider the sequence of numbers defined recursively by $t_1 = 1$ and for $n > 1$ by $t_n = 1 + t_{(n/2)}$ when n is even and by $t_n = 1/t_{(n-1)}$ when n is odd. Given that $t_n = 19/87$, the sum of the digits of n is

(A) 15   (B) 17   (C) 19   (D) 21   (E) 23

30. In the figure, $\triangle ABC$ has $\angle A = 45°$ and $\angle B = 30°$. A line DE, with D on AB and $\angle ADE = 60°$, divides $\triangle ABC$ into two pieces of equal area. (Note: the figure may not be accurate; perhaps E is on CB instead of AC.) The ratio AD/AB is

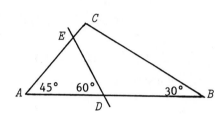

(A) $\dfrac{1}{\sqrt{2}}$ (B) $\dfrac{2}{2 + \sqrt{2}}$ (C) $\dfrac{1}{\sqrt{3}}$ (D) $\dfrac{1}{\sqrt[3]{6}}$ (E) $\dfrac{1}{\sqrt[4]{12}}$

1988 AHSME

1. $\sqrt{8} + \sqrt{18} =$

   (A) $\sqrt{26}$ (B) $2(\sqrt{2} + \sqrt{3})$ (C) 7   (D) $5\sqrt{2}$ (E) $2\sqrt{13}$

2. Triangles ABC and XYZ are similar, with A corresponding to X and B to Y. If $AB = 3$, $BC = 4$ and $XY = 5$, then YZ is

   (A) $3\frac{3}{4}$ (B) 6   (C) $6\frac{1}{4}$ (D) $6\frac{2}{3}$ (E) 8

3. Four rectangular paper strips of length 10 and width 1 are put flat on a table and overlap perpendicularly as shown. How much area of the table is covered?

   (A) 36   (B) 40   (C) 44   (D) 96
   (E) 100

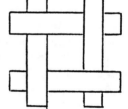

4. The slope of the line $\frac{x}{3} + \frac{y}{2} = 1$ is

   (A) $-\frac{3}{2}$ (B) $-\frac{2}{3}$ (C) $\frac{1}{3}$ (D) $\frac{2}{3}$ (E) $\frac{3}{2}$

5. If b and c are constants and

$$(x+2)(x+b) = x^2 + cx + 6,$$

   then c is

   (A) -5 (B) -3 (C) -1 (D) 3   (E) 5

6. A figure is an equiangular parallelogram if and only if it is a

   (A) rectangle   (B) regular polygon   (C) rhombus
   (D) square   (E) trapezoid

33

7. Estimate the time it takes to send 60 blocks of data over a communications channel if each block consists of 512 "chunks" and the channel can transmit 120 chunks per second.

(A) .04 seconds   (B) .4 seconds   (C) 4 seconds

(D) 4 minutes   (E) 4 hours

8. If $\dfrac{b}{a} = 2$ and $\dfrac{c}{b} = 3$, what is the ratio of $a+b$ to $b+c$?

(A) $\dfrac{1}{3}$ (B) $\dfrac{3}{8}$ (C) $\dfrac{3}{5}$ (D) $\dfrac{2}{3}$ (E) $\dfrac{3}{4}$

9. An $8' \times 10'$ table sits in the corner of a square room, as in Figure 1 below. The owners desire to move the table to the position shown in Figure 2. The side of the room is S feet. What is the smallest integer value of S for which the table can be moved as desired without tilting it or taking it apart?

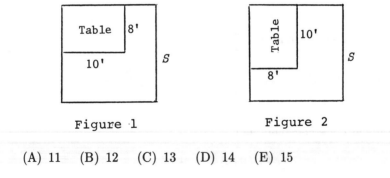

Figure 1   Figure 2

(A) 11   (B) 12   (C) 13   (D) 14   (E) 15

10. In an experiment, a scientific constant C is determined to be 2.43865 with an error of at most ± 0.00312. The experimenter wishes to announce a value for C in which every digit is significant. That is, whatever C is, the announced value must be the correct result when C is rounded to that number of digits. The most accurate value the experimenter can announce for C is

(A) 2.   (B) 2.4   (C) 2.43   (D) 2.44   (E) 2.439

11. On each horizontal line in the figure below, the five large dots indicate the populations of cities A, B, C, D and E in the year indicated. Which city had the greatest percentage increase in population from 1970 to 1980?

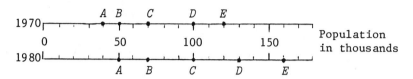

(A) A (B) B (C) C (D) D (E) E

12. Each integer 1 through 9 is written on a separate slip of paper and all nine slips are put into a hat. Jack picks one of these slips at random and puts it back. Then Jill picks a slip at random. Which digit is most likely to be the units digit of the **sum** of Jack's integer and Jill's integer?

(A) 0   (B) 1   (C) 8   (D) 9   (E) each digit is equally likely

13. If $\sin x = 3\cos x$ then what is $\sin x \cos x$?

(A) 1/6   (B) 1/5   (C) 2/9   (D) 1/4   (E) 3/10

14. For any real number a and positive integer k, define

$$\binom{a}{k} = \frac{a(a-1)(a-2)\cdots(a-(k-1))}{k(k-1)(k-2)\cdots(2)(1)}.$$

What is

$$\binom{-\frac{1}{2}}{100} \div \binom{\frac{1}{2}}{100} ?$$

(A) -199 (B) -197 (C) -1 (D) 197 (E) 199

15. If a and b are integers such that $x^2 - x - 1$ is a factor of $ax^3 + bx^2 + 1$, then b is

(A) -2 (B) -1 (C) 0 (D) 1 (E) 2

16. ABC and $A'B'C'$ are equilateral triangles with parallel sides and the same center, as in the figure. The distance between side BC and side $B'C'$ is $\frac{1}{6}$ the altitude of $\triangle ABC$. The ratio of the area of $\triangle A'B'C'$ to the area of $\triangle ABC$ is

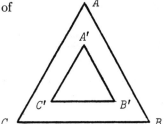

(A) $\dfrac{1}{36}$ (B) $\dfrac{1}{6}$ (C) $\dfrac{1}{4}$

(D) $\dfrac{\sqrt{3}}{4}$ (E) $\dfrac{9+8\sqrt{3}}{36}$

17. If $|x| + x + y = 10$ and $x + |y| - y = 12$, find $x + y$.

(A) -2 (B) 2 (C) $\dfrac{18}{5}$ (D) $\dfrac{22}{3}$ (E) 22

18. At the end of a professional bowling tournament, the top 5 bowlers have a play-off. First #5 bowls #4. The loser receives 5th prize and the winner bowls #3 in another game. The loser of this game receives 4th prize and the winner bowls #2. The loser of this game receives 3rd prize and the winner bowls #1. The winner of this game gets 1st prize and the loser gets 2nd prize. In how many orders can bowlers #1 through #5 receive the prizes?

(A) 10   (B) 16   (C) 24   (D) 120   (E) none of these

19. Simplify

$$\frac{bx(a^2x^2+2a^2y^2+b^2y^2) + ay(a^2x^2+2b^2x^2+b^2y^2)}{bx + ay}.$$

(A) $a^2x^2 + b^2y^2$ (B) $(ax+by)^2$ (C) $(ax+by)(bx+ay)$

(D) $2(a^2x^2+b^2y^2)$ (E) $(bx+ay)^2$

20. In one of the adjoining figures a square of side 2 is dissected into four pieces so that E and F are the midpoints of opposite sides and AG is perpendicular to BF. These four pieces can then be reassembled into a rectangle as shown in the second figure. The ratio of height to base, XY/YZ, in this rectangle is

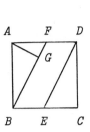
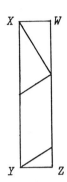

(A) 4   (B) $1 + 2\sqrt{3}$

(C) $2\sqrt{5}$ (D) $(8 + 4\sqrt{3})/3$

(E) 5

21. The complex number z satisfies $z + |z| = 2 + 8i$. What is $|z|^2$? Note: if $z = a + bi$, then $|z| = \sqrt{a^2 + b^2}$.

(A) 68   (B) 100   (C) 169   (D) 208   (E) 289

22. For how many integers x does a triangle with side lengths 10, 24 and x have all its angles acute?

(A) 4   (B) 5   (C) 6   (D) 7   (E) more than 7

23. The six edges of tetrahedron $ABCD$ measure 7, 13, 18, 27, 36 and 41 units. If the length of edge AB is 41, then the length of edge CD is

(A) 7   (B) 13   (C) 18   (D) 27   (E) 36

24. An isosceles trapezoid is circumscribed around a circle. The longer base of the trapezoid is 16, and one of the base angles is $\arcsin(.8)$. Find the area of the trapezoid.

(A) 72   (B) 75   (C) 80   (D) 90

(E) not uniquely determined

25. X, Y, and Z are pairwise disjoint sets of people. The average ages of people in the sets X, Y, Z, $X \cup Y$, $X \cup Z$ and $Y \cup Z$ are given in the table below.

Set	X	Y	Z	$X \cup Y$	$X \cup Z$	$Y \cup Z$
Average age of people in the set	37	23	41	29	39.5	33

Find the average age of the people in the set $X \cup Y \cup Z$.

(A) 33   (B) 33.5   (C) $33.6\overline{6}$ (D) $33.8\overline{33}$ (E) 34

26. Suppose that p and q are positive numbers for which

$$\log_9(p) = \log_{12}(q) = \log_{16}(p + q).$$

What is the value of q/p?

(A) 4/3   (B) $\frac{1}{2}(1+\sqrt{3})$ (C) 8/5   (D) $\frac{1}{2}(1+\sqrt{5})$ (E) 16/9

27. In the figure, $AB \perp BC$, $BC \perp CD$ and BC is tangent to the circle with center O and diameter AD. In which one of the following cases is the area of $ABCD$ an integer?

(A) $AB = 3$, $CD = 1$

(B) $AB = 5$, $CD = 2$

(C) $AB = 7$, $CD = 3$

(D) $AB = 9$, $CD = 4$

(E) $AB = 11$, $CD = 5$

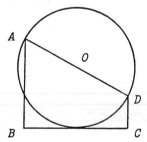

28. An unfair coin has probability p of coming up heads on a single toss. Let w be the probability that, in 5 independent tosses of this coin, heads come up exactly 3 times. If $w = 144/625$, then

(A) p must be 2/5   (B) p must be 3/5

(C) p must be greater than 3/5

(D) p is not uniquely determined

(E) there is no value of p for which $w = 144/625$

29. You plot weight (y) against height (x) for three of your friends and obtain the points (x_1, y_1), (x_2, y_2), (x_3, y_3). If

$$x_1 < x_2 < x_3 \quad \text{and} \quad x_3 - x_2 = x_2 - x_1,$$

which of the following is necessarily the slope of the line which best fits the data? "Best fits" means that the sum of the squares of the vertical distances from the data points to the line is smaller than for any other line.

(A) $\dfrac{y_3 - y_1}{x_3 - x_1}$ (B) $\dfrac{(y_2-y_1) - (y_3-y_2)}{x_3 - x_1}$ (C) $\dfrac{2y_3 - y_1 - y_2}{2x_3 - x_1 - x_2}$

(D) $\dfrac{y_2-y_1}{x_2-x_1} + \dfrac{y_3-y_2}{x_3-x_2}$ (E) none of these

30. Let $f(x) = 4x - x^2$. Given x_0, consider the sequence defined by $x_n = f(x_{n-1})$ for all $n \geq 1$. For how many real numbers x_0 will the sequence $x_0, x_1, x_2, \ldots$ take on only a finite number of different values?

(A) 0   (B) 1 or 2   (C) 3, 4, 5 or 6

(D) more than 6 but finitely many   (E) infinitely many

Dropped AHSME Problems

Early drafts of every exam contain extra problems. In part, this is so that the difficulty level and topic mix can be adjusted by deletions. But in part it is because, during the review process, some problems are seen to be flawed. Sometimes they are simply wrong, but more often, it turns out that they are not suitable for a multiple choice exam. They may be ambiguous in ways that cannot be overcome in brief wording. We may find it impossible to write good distractors – wrong answers that don't give a hint and can't be eliminated easily. It may turn out that it is easy to get the right answer for the wrong reason. The problem may involve too many subtleties or details for a timed format. Or, we may not have been able to give a complete solution with the mathematics studied in high schools. For whatever reasons, the Problems Committee frequently had to abandon problems whose key ideas we liked very much.

Below is a selection of dropped problems, often with an explanation. You may need to spend more time on them than on most AHSME problems. We do not recommend doing these problems under a timed format.

Some problems below were considered for the 1981 and 1982 AHSMEs, exams in Contest Problem Book IV. But since Book IV did not include dropped problems, and since we have the materials from which those exams were produced, we have included such problems.

Eventually (sometimes years later) most submitted problems that were not used were "released"; the author or others were free to use them elsewhere. So some of the problems below may have appeared elsewhere by now, probably in modified form. (If readers have seen any of these problems before, please let us know.)

The problems that follow are roughly in the chronological order they were considered by the Committee. They are not in order of difficulty.

1. If a is a real number, and

$$\sqrt{x^2 - 7ax + 10a^2} - \sqrt{x^2 + ax - 6a^2} = x - 2a,$$

then the sum of the real solutions for x is always

(A) $-4a/3$ (B) $8a/3$ (C) $14a/3$ (D) $8a$

(E) none of these

Remark. This problem was deleted for being too subtle; the original solution submitted by the proposer was wrong, and so were some of the replacement solutions. As with most radical equations, the subtlety involves extraneous roots. The Committee felt this is an important topic, since it gets at the logic behind all algebraic solution methods, but we couldn't devise a satisfactory version of this problem. For each version it was too easy to miss some case and get the wrong answer, or get the right answer for the wrong reason. Be sure to justify each step of your solution carefully!

2. Let $g(x) = (x-3)^2(x-1)^3(x+2)^2(x+3)^2$. If ϵ is a very small positive number, how many distinct real roots does $h(x) = g(x) + \epsilon$ have?

(A) 4   (B) 5   (C) 6   (D) 7   (E) 9

Remark. The Committee thought this problem addressed an important point about perturbation of roots of polynomials, but its statement relied on an understanding of how ϵ is used in higher mathematics. The following alternative wording was proposed: For some positive number b and all positive numbers c such that $0 < c < b$, the number of distinct real solutions to $(x-3)^2(x-1)^3(x+2)^2(x+3)^2 + c = 0$ is (same answers as before). But it was felt this wording was confusing in other ways, and made the problem sound artificial. There was also concern that our solution was incomplete. Complete justification that polynomials behave the way the solution claims is beyond secondary mathematics, but this behavior should be well known to strong secondary mathematics students.

3. If the coefficients a, b and c are all nonzero, then the roots of the equation $ax^2 + bx + c = 0$ are

(A) $\dfrac{2c}{-b \pm \sqrt{b^2 - 4ac}}$ (B) $\dfrac{2c}{-b \pm \sqrt{b^2 + 4ac}}$ (C) $\dfrac{b \pm \sqrt{b^2 - 4ac}}{2a}$

(D) $\dfrac{-b \pm \sqrt{b^2 + 4ac}}{2a}$ (E) none of these

4. Real numbers a, b, c, d and e, no two of which are equal, satisfy the equations

$$a - 2c + e = 0,$$
$$b - 2c + d = 0,$$
$$c - 2d + e = 0.$$

Then the largest and the smallest of the numbers a, b, c, d, e are both members of the set

(A) $\{a, b\}$ (B) $\{a, e\}$ (C) $\{b, d\}$ (D) $\{d, e\}$

(E) none of these sets

5. The discriminant of a quadratic equation with integer coefficients cannot be

(A) 23   (B) 24   (C) 25   (D) 28   (E) 33

6. How many ten-digit numbers (written in base ten) have this property: For $1 \le k \le 10$, the kth digit from the left is equal to the number of times the digit $k-1$ appears in the number.

(A) 0   (B) 1   (C) 2   (D) 3   (E) more than 3

7. A bus is traveling at constant speed, in the direction $30°$ north of east, along a straight road through flat open country. A runner is some distance off the road, due east of the present position of the bus. She can run 15 mph. She wishes to reach the road before or at the same time the bus reaches the same spot, so she can flag it down. This is possible, but if the bus were going any faster it would not be possible, no matter what point on the road she ran to. The speed of the bus in mph is

(A) 30   (B) $15\left(1 + \dfrac{\sqrt{3}}{2}\right)$ (C) $15\sqrt{3}$ (D) $10\sqrt{3}$

(E) cannot be determined

8. For all real numbers x and integers $n \geq 1$, define $f_1(x) = |1 - x|$ and $f_{n+1}(x) = |1 - f_n(x)|$. The equation $f_{1981}(x) = f_{1982}(x)$ has how many real solutions?

(A) 1981   (B) 1982   (C) 3962   (D) 3963   (E) none of these

9. In the adjoining diagram, circle O is inscribed in equilateral triangle ABC and segment DE is tangent to circle O, intersecting sides AB and AC at D and E, respectively. If DE is perpendicular to AC and $AE = 1$, then the side length of $\triangle ABC$ is

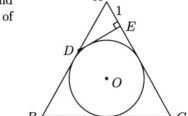

(A) $3\sqrt{3}$ (B) $3 + \sqrt{3}$

(C) $6 - \sqrt{3}$ (D) $(2+\sqrt{3})/2$

(E) $\sqrt{6} + \sqrt{3}$

10. How many different values of n are there such that n is a natural number and $n^2 - 440$ is a perfect square?

(A) 1   (B) 2   (C) 3   (D) 4   (E) infinitely many

11. Let $C(a)$ be the circle with center $(2a, 0)$ and radius a. Let P be the set of points contained in $C(a)$ for at least one a in the interval $0 \leq a \leq 2$. Then the area of P is

(A) 4π (B) $4\sqrt{3} + 4\pi$ (C) $4\sqrt{3} + \dfrac{4\pi}{3}$ (D) $8 + 2\pi$

(E) $4\sqrt{3} + \dfrac{8\pi}{3}$

Remark. This was deleted in part because the draft it appeared on had too much geometry, in part because some ideas might be too unfamiliar (e.g., the union of infinitely many figures, and circles of 0 radius). Also, participants might skip the problem because it isn't clear whether the circle $C(a)$ means just the set of boundary points or the interior as well (it turns out P is the same either way).

12. A candy manufacturer buys sugar each month. Between one month and the next, the price of sugar goes up by $p\%$, and the amount of money the manufacturer has for buying sugar goes up by $q\%$, where $0 < q < p$. The amount of sugar the manufacturer can buy

(A) increases by exactly $p - q$ %

(B) decreases by exactly $p - q$ %

(C) increases by more than $p - q$ %

(D) decreases by more than $p - q$ %

(E) decreases by less than $p - q$ %

Remark. The Problem Committee liked the idea of having an applied question on the combined effect of competing percentage changes, but we couldn't find a context where extraneous aspects might not confuse participants. For our solution to be correct, the quantities (in this case sugar and money) have to be "infinitely divisible" – you can take exactly any percentage you want. This is not true of money, and it is not true for sugar either if the producer insists on selling it, say, in 100 kg lots. Also, since the wording of the problem indicates that the answer is independent of the specific values of p and q, it is easy to pick simple values and find the answer by elimination. Finally, the right answer may give the wrong message; see the solution.

The following problem is about alternative ways to write an algebraic expression. It was dropped because it was too easy to eliminate the wrong answers by plugging in small values for the variable k. In general, it is very hard to write good multiple-choice problems about equivalent algebraic expressions, problems that can only be answered by really solving them. Below please look at the answers just to see the type of form involved, and then try to work the problem without looking at them any more.

13. For k a positive integer and n a real number, define

$$\binom{n}{k} = \frac{n(n-1)\cdots(n-k+1)}{k(k-1)\cdots 2\cdot 1}.$$

Then $\binom{1/2}{k}$ is equal to

(A) $\left(-\frac{1}{4}\right)^{k-1}\frac{1}{2k}\binom{2k-2}{k-1}$ (B) $\left(-\frac{1}{4}\right)^{k}\binom{2k}{k}$ (C) $\left(-\frac{1}{2}\right)^{k}\frac{1}{k+1}\binom{2k}{k}$

(D) $\left(-\frac{1}{4}\right)^{k}\frac{1}{k}\binom{2k-2}{k-1}$ (E) $\left(-\frac{1}{2}\right)^{k-1}\frac{(2k-3)!}{k!}$

14. A pocket calculator was used to determine the difference

$$\sqrt[3]{1000.000\,000\,030} \; - \; \sqrt[3]{1000.000\,000\,015}$$

and gave the answer 0.000000000, which was not accurate enough for subsequent use of the calculation! Which of the following is the closest approximation to the exact difference?

(A) $0.000\,000\,015$ (B) $0.000\,000\,005$ (C) $0.000\,000\,000\,15$

(D) $0.000\,000\,000\,05$ (E) $0.000\,000\,000\,001\,5$

15. 100 books were distributed among 30 people so that each person received at least one book. Let n be the largest number of people who received the same number of books each. Then the smallest possible value of n is

(A) 2   (B) 3   (C) 4   (D) 5   (E) 6

16. In the small hamlet of Abaze, two base systems are in common use. Also, everyone speaks the truth. One resident said, "26 people use my base, base 10, and only 22 people speak base 14." Another said, "Of the 25 residents, 13 are bilingual and 1 can't use either base." How many residents are there? (Use base 10, please!)

(A) 15   (B) 25   (C) 27   (D) 35   (E) 36

17. The largest acute angle satisfying the equation $\sin x + \cos x = 4/3$ is

(A) $\frac{1}{2}\sin^{-1}\frac{7}{9}$ (B) $\frac{1}{2}\cos^{-1}\frac{7}{9}$ (C) $\sin^{-1}\left(\frac{2\sqrt{2}}{3}\right) - \frac{\pi}{4}$

(D) $\cos^{-1}\left(\frac{2\sqrt{2}}{3}\right) + \frac{\pi}{4}$ (E) none of these

Remark. This problem was one of several attempts to get at some interesting ideas about inverse trigonometric functions. Trigonometric equations have many solutions, but inverse trig functions, being functions (unique output), can express only one answer at a time. We wanted students to think about the implications of this fact.

However, we dropped these attempts because there are so many correct ways to express an angle using inverse trig functions. Unless you solved for the angle the same way we did, you would get an answer that looks completely different from *any* of the answers offered. Thus you could do very good, correct work and have no idea which answer to choose. The original version of this problem made the situation even worse: it merely asked for *some* correct angle, so you could also find another correct angle and again get no credit.

There were still other difficulties. The notation for inverse trigonometric functions is not standardized. Some books use arcsin, some $\sin^{-1}$, some Sin^{-1}. In some places Sin^{-1} stands for the inverse *function* (that is, the range of sin is restricted so that it has an inverse), and $\sin^{-1}$ stands for the inverse relation (that is, $\sin^{-1}\frac{1}{2}$ is the set of *all* angles whose sin is $\frac{1}{2}$). Finally, even if you understood $\sin^{-1}$ to be a function, it might be unreasonable to expect you to remember its range, or the range might not be standardized in all books.

In writing this problem, we assumed that $\sin^{-1}$ is a function with range $[-\frac{\pi}{2}, \frac{\pi}{2}]$ and $\cos^{-1}$ is a function with range $[0, \pi]$.

18. If $a + b = 1$ and $a^2 + b^2 = 2$ then $a^3 + b^3$ equals

(A) 4   (B) $3\frac{1}{2}$ (C) 3   (D) $2\frac{1}{2}$ (E) 2

19. Let r be the radius of the largest circle which is tangent to the parabola $y = x^2$ at $x = 0$ and which lies entirely on or inside (that is, above) that parabola. Then

(A) $r = 1/4$ (B) $r = 1/3$ (C) $r = 1/2$ (D) $r = 1$

(E) No such circle is the largest

20. If the function $y = \log_x(x+1)$ is plotted for all real values of x for which it is defined, the graph looks like

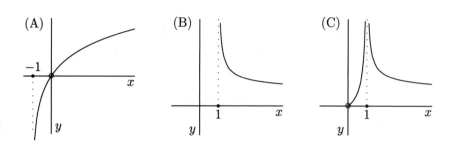

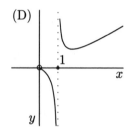

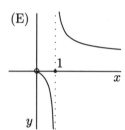

21. The domain of $\log_{1/2}\left(\log_2\left(\log_{1/2}x\right)\right)$ is

(A) $\{x : x > 0\}$ (B) $\{x : x > 1/2\}$ (C) $\{x : 0 < x < 1\}$

(D) $\{x : 0 < x < 1/2\}$ (E) the empty set

22. An equal-arm balance used in a school chemistry lab is defective. One arm is 5% longer than the other. A student attempts to measure 420 grams of a certain chemical by placing calibrated weights totaling 210 grams in pan A and balances the weights with W_1 grams of the chemical in pan B. He then places the calibrated weights totaling 210 grams in pan B and balances the weights with W_2 grams of the chemical in pan A. Then $W_1 + W_2$ is

(A) 420 grams   (B) 441 grams   (C) 420.5 grams

(D) 399 grams   (E) none of these

Remark. Mathematically, this is a very nice problem about the idea of compensating for error through symmetrization. However, the problem has two difficulties. First, many students may have no familiarity with such old-fashioned mechanical balance scales. Second, readers must know the appropriate physical principle: objects will balance if $d_1 W_1 = d_2 W_2$, where W_1, W_2 are the weights of the masses on opposite arms of the balance, and d_1, d_2 are their respective distances from the balance point.

23. A circle passes through vertex C of equilateral triangle ABC and is tangent to side AB at point F between A and B. The circle meets AC and BC respectively at D and E. If $AF/FB = p$, then AD/BE is

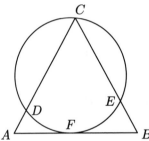

(A) p (B) p^2

(C) $\sqrt{p}$ (D) $\sqrt{3}\left(p + \dfrac{1}{p}\right)$

(E) not uniquely determined

24. Let $\lfloor x \rfloor$ be the greatest integer $\le x$. Let r be a random number strictly between 0 and 1. Then which expression below gives an integer from 0 to 10, inclusive, with each integer equally likely?

(A) $\lfloor 10r \rfloor$ (B) $10\lfloor r \rfloor$ (C) $\lfloor 10r + .5 \rfloor$ (D) $\lfloor 11r \rfloor$

(E) $11\lfloor r \rfloor$

25. A circle of radius r goes through two neighboring vertices of a square and is tangent to the side of the square opposite these vertices. In terms of r, the area of the square is

(A) $\frac{8}{5}r^2$ (B) $2r^2$ (C) $\frac{64}{25}r^2$ (D) $\frac{8}{3}r^2$ (E) πr^2

26. When a fair coin is tossed n times, the probability is approximately 95% that the total number of heads will be within $\sqrt{n}$ of $n/2$. Use this fact to determine how many times a fair coin must be tossed if one wants a probability of approximately 95% that between 40% and 60% of the tosses will be heads.

(A) 5   (B) 10   (C) 25   (D) 50   (E) 100

Remark. This problem, proposed for the 1988 exam, was an attempt to get some ideas about sampling accuracy onto the AHSME, in an continuing effort to indicate that statistical ideas are important. However, there was already another coin toss problem under consideration and the committee preferred it. We worried that this problem introduced too many ideas that would be new to most participants and that therefore it might not provide an accurate measure of ability. Also, the use of "approximately", though necessary, made the problem vague in a way that might make more than one answer defensible.

27. Which of the following is a factor of $ab(c^2 + d^2) + cd(a^2 + b^2)$?

(A) ab (B) $c^2 + d^2$ (C) $ab + cd$ (D) $ac + bd$
(E) none of these

AHSME Answer Key

1983 AHSME

1. E	7. B	13. E	19. A	25. B
2. B	8. A	14. E	20. C	26. D
3. A	9. D	15. C	21. D	27. E
4. D	10. D	16. D	22. B	28. C
5. D	11. B	17. C	23. A	29. C
6. C	12. E	18. B	24. E	30. C

1984 AHSME

1. C	7. C	13. A	19. D	25. D
2. B	8. D	14. A	20. C	26. B
3. C	9. D	15. A	21. C	27. C
4. B	10. B	16. E	22. B	28. C
5. D	11. A	17. C	23. D	29. A
6. B	12. B	18. E	24. E	30. B

1985 AHSME

1. A	7. E	13. E	19. A	25. B
2. E	8. E	14. C	20. D	26. E
3. D	9. B	15. E	21. C	27. C
4. C	10. E	16. B	22. D	28. B
5. E	11. B	17. B	23. C	29. C
6. B	12. D	18. D	24. C	30. E

1986 AHSME

1. B	7. E	13. E	19. A	25. B
2. A	8. E	14. D	20. E	26. C
3. D	9. C	15. C	21. B	27. C
4. B	10. E	16. C	22. C	28. C
5. A	11. B	17. B	23. E	29. B
6. C	12. B	18. E	24. D	30. B

1987 AHSME

1. D	7. C	13. A	19. B	25. C
2. E	8. C	14. B	20. A	26. B
3. C	9. B	15. D	21. B	27. B
4. B	10. D	16. D	22. C	28. D
5. B	11. E	17. E	23. D	29. A
6. A	12. D	18. D	24. B	30. E

1988 AHSME

1. D	7. D	13. E	19. B	25. E
2. D	8. B	14. A	20. E	26. D
3. A	9. C	15. A	21. E	27. D
4. B	10. D	16. C	22. A	28. D
5. E	11. C	17. C	23. B	29. A
6. A	12. A	18. B	24. C	30. E

Dropped AHSME

1. E	7. A	13. A	19. C	25. C
2. B	8. C	14. D	20. E	26. E
3. A	9. B	15. E	21. D	27. D
4. B	10. D	16. D	22. C	
5. A	11. E	17. D	23. B	
6. D	12. E	18. D	24. D	

Response Frequency Tables

The following tables give the distribution of answers on the AHSME for all Honor Roll students (those eligible to take the AIME). Data for all students is not available since the AHSME was not centrally scored.

Answers are given in percents. For instance, 0.65% of Honor Roll students gave answer B to problem 1 on the 1983 AHSME. Boldface indicates the correct answer, so E was the correct answer to problem 1 in 1983. The last column (Blank) is the percent who didn't answer the question.

From these tables readers can get an idea of how hard other very good students found these problems. Generally, the larger the percent who didn't answer a problem, the harder the problem. (This conclusion may be less accurate for problems toward the end of each exam since many students simply didn't get to them within the time limit.) It is especially instructive to look at "popular" wrong answers. Any wrong answer that received at least 5% response probably indicates a common misconception or computational mistake. If you picked one of these incorrect answers, analyze carefully where you went wrong. Many people feel they learn more from their failures/errors than from their successes.

For instance, the 10% of Honor Roll students who answered C on problem 10 in 1983 probably confused radius and diameter in the statement of the problem (a careless, time-pressure mistake). But for problem 26, the 31.48% who answered B (*twice* the number who gave the correct answer D!) almost surely made a much more substantial error: they assumed that probability events are automatically independent.

1983 AHSME

Prob.	A	B	C	D	E	Blank
1	2.78	0.65	1.48	1.39	**93.43**	0.28
2	1.30	**97.22**	0.19	0.37	0.09	0.83
3	**98.15**	0.46	0.56	0.19	0.00	0.65
4	2.59	0.65	0.19	**94.44**	1.20	0.93
5	0.09	0.65	4.44	**93.24**	0.09	1.48
6	0.56	0.28	**97.41**	0.65	0.28	0.83
7	0.09	**97.41**	0.56	0.37	0.37	1.20
8	**96.48**	0.37	0.09	0.19	0.37	2.50
9	0.19	0.83	0.28	**96.67**	0.19	1.85
10	0.56	0.65	10.00	**83.15**	0.93	4.72
11	0.83	**85.09**	1.11	4.26	1.30	7.41
12	0.93	1.57	0.19	0.56	**91.39**	5.37
13	1.20	0.37	0.37	1.11	**89.35**	7.59
14	2.96	3.33	0.46	3.52	**83.06**	6.67
15	8.06	1.11	**82.78**	2.78	2.04	3.24
16	2.13	1.85	4.17	**77.22**	7.87	6.76
17	3.89	3.24	**73.89**	9.63	1.02	8.33
18	3.06	**65.09**	1.11	1.85	10.37	18.52
19	**53.06**	6.48	4.07	1.94	1.67	32.78
20	1.20	7.31	**62.78**	3.43	1.30	23.98
21	10.28	2.78	1.76	**63.98**	2.31	18.89
22	1.02	**18.80**	1.67	4.44	13.61	60.46
23	**47.41**	3.52	13.80	1.02	1.57	32.69
24	12.13	7.41	6.20	1.11	**34.35**	38.80
25	2.87	**29.63**	3.70	2.41	3.70	57.69
26	3.80	31.48	5.09	**15.93**	6.39	37.31
27	17.87	1.39	2.69	2.87	**21.85**	53.33
28	0.93	4.54	**8.89**	3.06	11.39	71.20
29	2.41	3.43	**10.19**	1.39	1.39	81.20
30	1.11	2.04	**18.98**	6.57	9.54	61.76

1984 AHSME

Prob.	A	B	C	D	E	Blank
1	0.00	1.16	**96.23**	0.58	2.03	0.00
2	0.29	**99.13**	0.29	0.00	0.29	0.00
3	0.29	0.29	**87.86**	0.00	11.56	0.00
4	1.16	**89.31**	4.62	0.58	0.87	3.47
5	0.00	0.29	0.87	**96.82**	1.16	0.87
6	2.02	**95.68**	1.15	0.86	0.29	0.00
7	1.16	0.58	**96.82**	0.29	0.29	0.87
8	0.00	0.00	2.31	**95.66**	1.45	0.58
9	0.29	0.58	1.15	**87.03**	2.59	8.36
10	1.45	**96.53**	0.29	0.29	0.58	0.87
11	**80.98**	1.15	1.44	1.73	13.54	1.15
12	3.19	**74.20**	0.87	14.49	4.06	3.19
13	**83.04**	1.46	1.17	1.46	0.88	11.99
14	**85.47**	1.74	9.59	0.58	1.16	1.45
15	**92.40**	0.58	2.92	0.29	0.29	3.51
16	7.78	2.31	3.75	1.73	**64.84**	19.60
17	1.16	0.87	**84.68**	0.00	1.73	11.56
18	2.31	2.31	31.70	3.17	**55.91**	4.61
19	0.87	7.23	4.62	**47.11**	2.60	37.57
20	1.45	5.49	**65.90**	3.18	21.39	2.60
21	11.24	15.27	**42.36**	3.17	5.48	22.48
22	9.51	**55.91**	3.46	3.75	2.02	25.36
23	0.58	4.32	4.61	**40.35**	3.75	46.40
24	2.88	3.17	4.32	4.90	**63.40**	21.33
25	1.73	1.73	1.44	**66.86**	6.92	21.33
26	2.02	**30.92**	2.31	4.05	7.80	52.89
27	4.91	2.31	**15.90**	5.20	3.76	67.92
28	5.20	4.05	**38.73**	6.94	4.91	40.17
29	**17.63**	9.83	2.89	1.73	1.73	66.18
30	2.31	**6.36**	1.73	1.45	5.78	82.37

1985 AHSME

Prob.	A	B	C	D	E	Blank
1	**99.65**	.00	.00	.00	.35	.00
2	.00	.17	2.95	1.39	**95.49**	.00
3	.00	.00	.00	**100.00**	.00	.00
4	.35	.00	**96.70**	.52	1.74	.69
5	.17	.00	.00	.52	**99.31**	.00
6	.00	**84.20**	.69	1.22	13.89	.00
7	.17	.52	1.91	5.03	**91.67**	.69
8	1.04	6.42	.69	3.65	**85.94**	2.26
9	2.08	**91.15**	2.60	2.26	.35	1.56
10	8.51	11.46	1.91	.69	**71.88**	5.56
11	2.78	**81.77**	12.85	.69	.17	1.74
12	.87	.52	1.04	**93.06**	1.74	2.78
13	.00	.17	.00	.52	**99.31**	.00
14	8.68	21.35	**53.65**	4.34	2.43	9.55
15	3.13	.69	2.26	3.47	**83.51**	6.94
16	1.91	**35.94**	3.30	3.65	8.85	46.35
17	2.43	**63.89**	1.74	1.04	1.91	28.99
18	1.56	.69	.52	**91.67**	.87	4.69
19	**48.44**	13.19	6.42	10.59	4.86	16.49
20	.17	2.95	3.13	**83.33**	5.73	4.69
21	9.55	37.15	**39.24**	2.08	3.65	8.33
22	.69	.35	1.39	**79.69**	5.03	12.85
23	2.08	1.74	**55.73**	1.74	1.04	37.67
24	2.26	2.78	**53.65**	2.78	5.56	32.99
25	5.03	**53.99**	3.30	1.04	1.04	35.59
26	1.56	1.56	29.86	5.73	**26.74**	34.55
27	.69	3.30	**52.26**	3.30	2.43	38.02
28	1.74	**9.55**	1.91	2.08	10.07	74.65
29	1.91	3.99	**32.29**	3.82	1.39	56.60
30	11.81	7.99	11.98	3.99	**9.90**	54.34

1986 AHSME

Prob.	A	B	C	D	E	Blank
1	.04	**97.99**	.39	.07	1.31	.21
2	**99.12**	.04	.28	.00	.46	.11
3	.14	.14	.28	**99.26**	.14	.04
4	1.59	**97.00**	.18	.07	.64	.53
5	**96.68**	.04	.07	.25	.14	2.82
6	.18	.25	**98.38**	.11	.21	.88
7	2.26	13.03	.81	.21	**80.44**	3.25
8	1.80	.78	5.82	2.05	**84.50**	5.05
9	.11	.46	**94.03**	.18	.14	5.08
10	3.18	1.94	8.51	1.52	**68.69**	16.17
11	.56	**57.89**	.85	2.54	.42	37.73
12	.74	**83.48**	.56	.28	4.69	10.24
13	2.68	.78	2.40	.42	**57.68**	36.04
14	4.94	.67	.53	**86.97**	.92	5.97
15	.14	3.71	**86.02**	.85	2.51	6.78
16	1.80	.81	**64.88**	.32	.14	32.05
17	.74	**66.64**	4.73	.42	1.06	26.40
18	.28	16.38	.95	.28	**46.63**	35.47
19	**66.15**	1.06	1.38	.95	1.13	29.33
20	2.97	2.15	1.16	15.21	**33.89**	44.62
21	6.04	**21.39**	2.05	.21	.56	69.75
22	1.02	2.26	**13.87**	.53	12.28	70.03
23	3.53	1.98	.71	3.92	**22.45**	67.42
24	.21	.85	.49	**5.51**	3.74	89.20
25	.88	**21.25**	1.94	4.73	5.40	65.80
26	1.20	5.22	**5.61**	.39	6.21	81.36
27	1.52	.28	**11.79**	.64	.56	85.21
28	1.62	2.44	**1.09**	2.05	.18	92.62
29	.53	**2.15**	.74	1.24	2.40	92.94
30	2.40	**3.25**	.53	.64	1.55	91.63

1987 AHSME

Prob.	A	B	C	D	E	Blank
1	0.10	0.02	0.02	**98.79**	1.07	0.00
2	0.07	0.02	0.94	0.00	**98.91**	0.05
3	0 15	2.40	**88.65**	4.81	3.07	0.92
4	0 07	**99.11**	0.35	0.00	0.22	0.25
5	0.64	**89.40**	0.22	0.22	0.07	9.44
6	**96.43**	0.02	0.37	0.10	1.02	2.06
7	0.15	0.00	**99.13**	0.10	0.32	0.30
8	1.71	0.67	**78.13**	2.38	16.10	1.02
9	0.17	**88.78**	7.48	0.15	0.45	2.97
10	16.92	12.53	1.88	**51.77**	2.97	13.92
11	0.20	4.95	2.40	3.42	**82.12**	6.91
12	0.05	0.32	0.12	**98.12**	0.05	1.34
13	**59.92**	1.11	1.39	6.42	0.82	30.34
14	2.11	**40.97**	3.10	1.34	12.61	39.88
15	1.51	0.79	0.22	**63.09**	0.77	33.61
16	0.20	0.42	0.25	**61.43**	1.44	36.26
17	1.41	2.48	0.22	1.11	**85.61**	9.17
18	0.27	0.10	0.25	**44.17**	0.12	55.09
19	14.09	**24.87**	1.11	0.40	1.07	58.46
20	**35.27**	0.32	0.15	1.36	1.63	61.26
21	0.37	**70.42**	0.40	1.21	0.30	27.30
22	0.20	1.07	**51.03**	1.34	0.74	45.63
23	0.30	0.20	0.57	**9.29**	1.07	88.58
24	1.09	**19.02**	1.83	0.45	1.54	76.07
25	1.26	0.42	**5.25**	2.45	8.45	82.16
26	0.54	**42.14**	2.38	3.15	0.22	51.57
27	0 27	**12.36**	0.57	22.99	0.47	63.34
28	0 77	0.37	1.76	**3.99**	0.20	92.92
29	**4.01**	0.54	0.20	0.22	0.22	94.80
30	0.22	0.15	0.69	0.12	**1.09**	97.72

1988 AHSME

Prob.	A	B	C	D	E	Blank
1	0.04	1.65	0.00	**98.24**	0.00	0.07
2	0.05	0.01	0.00	**99.77**	0.03	0.14
3	**99.54**	0.11	0.05	0.23	0.00	0.07
4	0.38	**98.98**	0.03	0.40	0.04	0.18
5	0.11	0.00	0.05	0.40	**98.54**	0.90
6	**91.30**	1.54	0.96	3.02	0.07	3.11
7	0.01	0.04	0.68	**97.84**	0.82	0.60
8	0.25	**96.50**	0.16	0.11	0.92	2.06
9	0.40	1.50	**78.83**	2.47	0.60	16.20
10	0.07	9.59	0.51	**82.32**	1.30	6.21
11	0.87	1.65	**95.22**	0.20	1.54	0.51
12	**70.95**	1.00	2.08	4.51	10.91	10.56
13	0.18	0.04	0.26	0.12	**64.19**	35.21
14	**17.53**	0.79	2.35	0.14	1.95	23.23
15	**61.00**	1.20	3.43	0.72	1.82	31.83
16	0.60	0.59	**45.70**	0.93	0.70	51.48
17	7.44	0.66	**39.87**	2.80	2.04	47.19
18	0.68	**46.97**	1.42	7.31	5.15	38.46
19	.026	**40.51**	0.63	0.40	0.81	57.40
20	3.20	0.34	8.06	0.40	**52.38**	35.62
21	1.15	0.20	0.16	0.16	**29.34**	68.98
22	**36.91**	5.08	1.23	2.12	6.88	47.78
23	3.89	**15.16**	0.96	0.31	0.49	79.18
24	0.55	0.07	**9.32**	0.16	3.44	86.46
25	0.15	0.44	1.97	3.07	**18.70**	75.67
26	1.58	0.72	0.11	**1.68**	0.22	95.68
27	0.96	1.31	1.15	**10.91**	0.85	84.82
28	4.77	0.27	2.47	**3.51**	1.52	87.46
29	**3.58**	0.48	0.61	0.22	0.90	94.21
30	0.63	1.16	3.39	0.53	**3.73**	90.56

1983 AHSME Solutions

1. (E) Solving each equation for y^2 gives

$$y^2 = \frac{x}{2} = \frac{x^2}{256}.$$

Thus

$$x = \frac{256}{2} = 128 \quad \text{since } x \neq 0.$$

2. (B) The points in question are the points of intersection of the original circle C and the circle of radius $3\,\text{cm}$ around P. Two distinct circles intersect in at most 2 points.

3. (A) Not both p and q are odd, since then r would be an even prime greater than 2, which is impossible. Thus one of p and q is 2. Since $1 < p < q$, p is 2.

4. (D) Draw in lines BF, BE and BD. There are now 4 equilateral triangles with side length 1. (For instance, $\triangle FAB$ is equilateral because $AF = AB = 1$ and $\angle A = 60°$.) Thus the total area is

$$4\,\frac{1^2\sqrt{3}}{4} = \sqrt{3}.$$

5. (D) In the figure, $\sin A = \dfrac{BC}{AB} = \dfrac{2}{3}$.
So for some $x > 0$
$BC = 2x$, $AB = 3x$ and

$$AC = \sqrt{(AB)^2 - (BC)^2} = \sqrt{5}\,x.$$

Thus $\tan B = \dfrac{AC}{BC} = \dfrac{\sqrt{5}}{2}$.

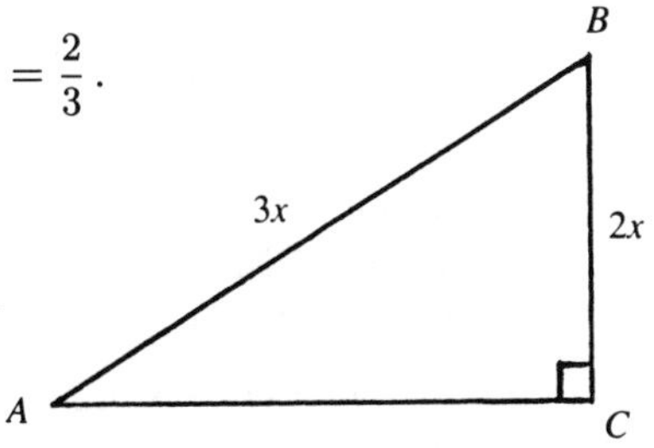

59

6. (C) By definition, a polynomial (in x) is an expression of the form

$$a_n x^n + a_{n-1} x^{n-1} + \cdots + a_0,$$

where the a's are constants and $a_n \neq 0$. The degree is defined to be n, the highest power of x. Since

$$x^5 \left(x + \frac{1}{x}\right)\left(1 + \frac{2}{x} + \frac{3}{x^2}\right) = x^2(x^2+1)(x^2+2x+3),$$

one sees without actually multiplying further (if one adds the degrees of the factors on the right) that the product is a polynomial and has degree 6.

7. (B) Let x be the list price in dollars. Then

$$.1(x-10) = .2(x-20)$$
$$x - 10 = 2x - 40$$
$$x = 30.$$

8. (A)
$$f(-x) = \frac{-x+1}{-x-1} = \frac{x-1}{x+1} = \frac{1}{\left(\dfrac{x+1}{x-1}\right)} = \frac{1}{f(x)}.$$

9. (D) Let w be the number of women, m the number of men. From the given, $w = 11x$ and $m = 10x$ for some x; also $34w$ is the sum of the ages of the women and $32m$ is the sum for the men. Thus the average age for all is.

$$\frac{34(11x) + 32(10x)}{11x + 10x} = \frac{34 \cdot 11 + 32 \cdot 10}{21} = \frac{694}{21} = 33\tfrac{1}{21}.$$

Note. Except for the leftmost expression in the previous line, the computation of the average is the same computation one would do if there were *exactly* 11 women and 10 men.

Remark. Many people think that when an average is asked for, and two numbers x and y are mentioned, the answer must be $(x+y)/2$. For this problem, that approach leads to answer (C). But averages can be *weighted*, as in this problem, and so (C) is wrong.

10. (D) In the figure, $\angle AEB = 90°$ since
    AB is a diameter. Thus AE is an alti-
    tude of equilateral $\triangle ABC$. It follows
    that $\triangle ABE$ is a 30°-60°-90° triangle
    and $AE = AB\dfrac{\sqrt{3}}{2} = \sqrt{3}$.

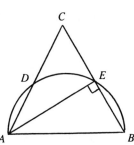

11. (B) Let $w = x-y$. Then the given expression is

$$\sin w \cos y + \cos w \sin y = \sin(w+y) = \sin x.$$

12. (E)
$$\log_7\big(\log_3(\log_2 x)\big) = 0 \implies \log_3(\log_2 x) = 1$$
$$\implies \log_2 x = 3 \implies x = 2^3 = 8.$$

So $x^{-1/2} = \dfrac{1}{\sqrt{8}} = \dfrac{1}{2\sqrt{2}}$.

13. (E) Observe that $abc = x^2y^2z^2 = x^2c^2$, so $x^2 = ab/c$. Likewise,
$y^2 = ac/b$ and $z^2 = bc/a$. So

$$x^2 + y^2 + z^2 = \frac{ab}{c} + \frac{ac}{b} + \frac{bc}{a} = \frac{(ab)^2 + (ac)^2 + (bc)^2}{abc}.$$

14. (E) Consider the first few powers of 3, 7 and 13:

3	7	13
9	49	169
27	343	...7
81	...1	1
243	7	3

Clearly, the units digits in each case go through a cycle of length 4, with the units digit being 1 if the power is a multiple of 4. Let $u(n)$ be the units digit of n. Since 4 divides 1000,

$$u(3^{1001}) = u(3^1) = 3,$$
$$u(7^{1002}) = u(7^2) = 9,$$
$$u(13^{1003}) = u(13^3) = 7.$$

So $u(3^{1001} 7^{1002} 13^{1003}) = u(3 \cdot 9 \cdot 7) = 9$.

Note. This solution can be expressed much more briefly using congruences.

OR

Any power of either $7 \cdot 13 = 91$ or $3^4 = 81$ has a units digit of 1. Thus $3^{1001} 7^{1002} 13^{1003} = 3 \cdot 13 \cdot 81^{250} 91^{1002}$, which clearly has a units digit of 9.

15. (C) A total of 6 can be achieved by 7 equally likely ordered triples of draws:

$$(1, 2, 3), \quad (1, 3, 2), \quad (2, 1, 3), \quad (2, 3, 1),$$
$$(3, 1, 2), \quad (3, 2, 1), \quad (2, 2, 2).$$

Therefore the answer is $1/7$.

16. (D) Look at the first 1983 digits, letting z denote the 1983${}^{\text{rd}}$digit.We may break this string of digits into three segments:

$$\underbrace{.123456789}_{A} \; \underbrace{1011 \ldots 9899}_{B} \; \underbrace{100101 \ldots z}_{C}.$$

There are 9 digits in A, $2 \cdot 90 = 180$ in B, hence $1983 - 189 = 1794$ in C. Dividing 1794 by 3 we get 598 with remainder 0. Thus C consists of the first 598 3-digit integers. Since the first 3-digit integer is 100 (not 101 or 001), the 598${}^{\text{th}}$ 3-digit integer is 697. Thus $z = 7$.

17. (C)   Write F as $a + bi$, where we see from the diagram that $a, b > 0$ and $a^2 + b^2 > 1$. Since

$$\frac{1}{a+bi} = \frac{a-bi}{a^2+b^2} = \frac{a}{a^2+b^2} - \frac{b}{a^2+b^2}\,i,$$

we see that the reciprocal of F is in quadrant IV, since the real part on the right is positive and the coefficient of the imaginary part is negative. Also, the magnitude of the reciprocal is

$$\frac{1}{a^2+b^2}\sqrt{a^2+(-b)^2} = \frac{1}{\sqrt{a^2+b^2}} < 1.$$

Thus the only possibility is point C.

OR

For any complex number $z \neq 0$, the argument (standard reference angle) of $1/z$ is the negative of the argument of z, and the modulus (magnitude) of $1/z$ is the reciprocal of the modulus of z. (Quick proof: if $z = re^{i\theta}$, then $\dfrac{1}{z} = \dfrac{1}{re^{i\theta}} = \dfrac{1}{r}e^{i(-\theta)}$.) Applying this to the point F, its reciprocal must be in quadrant IV, inside the unit circle. The only possibility is point C.

18. (B)   Since $f(x)$ is a polynomial and $f(x^2+1)$ has degree 4, then $f(x)$ has degree 2. That is, $f(x) = ax^2+bx+c$ for some constants a, b, c, and

$$f(x^2+1) = x^4 + 5x^2 + 3 = a(x^2+1)^2 + b(x^2+1) + c$$
$$= ax^4 + (2a+b)x^2 + (a+b+c).$$

Since two polynomials are equal if and only if the coefficients of corresponding terms are equal, we have

$$a = 1, \quad 2a + b = 5 \quad \text{and} \quad a + b + c = 3.$$

Solving these gives

$$f(x) = x^2 + 3x - 1.$$

Thus

$$f(x^2-1) = (x^2-1)^2 + 3(x^2-1) - 1 = x^4 + x^2 - 3.$$

OR

Rewrite x^4+5x^2+3 in terms of powers of $x^2 + 1$:

$$x^4 + 5x^2 + 3 = (x^4+2x^2+1) + (3x^2+3) - 1$$
$$= (x^2+1)^2 + 3(x^2+1) - 1.$$

Thus setting $w = x^2+1$, we have $f(w) = w^2+3w-1$.

(Actually, since x is real, w as defined can only take on values equal to or greater than 1. Thus so far we have shown only that $f(w) = w^2+3w-1$ when $w \geq 1$. However, since $f(w)$ is given to be a polynomial, and since it is identical to the polynomial w^2+3w-1 for infinitely many w, it follows that they are identical for all w.)

Therefore, as before, for all x

$$f(x^2-1) = (x^2-1)^2 + 3(x^2-1) - 1 = x^4 + x^2 - 3.$$

19. (A)   Let $AD = y$. Since AD bisects $\angle BAC$, we have $\dfrac{DB}{CD} = \dfrac{AB}{BC} = 2$;   so we may set $CD = x$, $DB = 2x$ as in the figure. Applying the Law of Cosines to $\triangle CAD$ and $\triangle DAB$, we have

$$x^2 = 3^2 + y^2 - 3y,$$
$$(2x)^2 = 6^2 + y^2 - 6y.$$

Subtracting 4 times the first equation from the second yields $0 = -3y^2+6y = -3y(y-2)$. Since $y \neq 0$, $y = 2$.

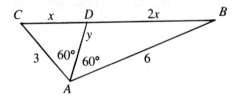

OR

Extend CA to E so that $BE \parallel DA$ as in the new figure. Then $\triangle ABE$ is equilateral: $\angle BEA = \angle DAC$ by corresponding angles, $\angle ABE = \angle BAD$ by alternate interior angles, and $\angle EAB = 180° - 120°$. Since $\triangle BEC \sim \triangle DAC$, we have $\dfrac{DA}{BE} = \dfrac{CA}{CE}$, or $\dfrac{DA}{6} = \dfrac{3}{9}$. So $DA = 2$.

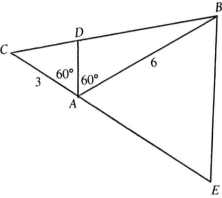

20. (C) By the relationship between the roots and the coefficients of a quadratic equation, it follows that

$$p = \tan\alpha + \tan\beta, \quad q = \tan\alpha\tan\beta,$$
$$r = \cot\alpha + \cot\beta, \quad s = \cot\alpha\cot\beta.$$

Since

$$\cot\alpha + \cot\beta = \frac{1}{\tan\alpha} + \frac{1}{\tan\beta} = \frac{\tan\alpha + \tan\beta}{\tan\alpha\tan\beta},$$

and

$$\cot\alpha\cot\beta = \frac{1}{\tan\alpha\tan\beta}$$

the equalities $r = p/q$ and $s = 1/q$ follow. Thus $rs = p/q^2$.

OR

The roots are reciprocals, and in general, if $x^2 - px + q = 0$ has roots m and n, then $qx^2 - px + 1 = 0$ has roots $1/m$ and $1/n$. Dividing the last equation by q shows that

$$x^2 - \frac{p}{q}x + \frac{1}{q} = 0$$

also has roots $1/m$ and $1/n$. So $r = p/q$, $s = 1/q$ and $rs = p/q^2$.

21. (D) Since

$$10^2 = 100 > 99 = (3\sqrt{11})^2,$$
$$18^2 = 324 < 325 = (5\sqrt{13})^2,$$
$$51^2 = 2601 > 2600 = (10\sqrt{26})^2,$$

only (A) and (D) are positive numbers. Moreover, since $a - b = (a^2 - b^2)/(a+b)$,

$$10 - 3\sqrt{11} = \frac{1}{10 + 3\sqrt{11}} \approx \frac{1}{20}$$

and

$$51 - 10\sqrt{26} = \frac{1}{51 + 10\sqrt{26}} \approx \frac{1}{102}.$$

So (D) is the answer.

Remark. This problem can be done almost as quickly, and with much less thought, using a calculator. Calculators were not allowed on the AHSME then; they are now. This problem could be changed to make ordinary calculators useless by making the distractors smaller than the roundoff errors of these machines. (See #14 in the Dropped AHSME Problems). However, the problem then becomes ungainly.

22. (B) We must describe geometrically those (a, b) for which the equation

$$x^2 + 2bx + 1 = 2a(x+b),$$

or equivalently, the equation

$$x^2 + 2(b-a)x + (1-2ab) = 0,$$

has no real root for x. Since a quadratic equation $Ax^2 + Bx + C = 0$ has no real root if and only if its discriminant $B^2 - 4AC$ is negative, S is the set of (a, b) for which

$$[2(b-a)]^2 - 4(1-2ab) < 0,$$
$$4(a^2 - 2ab + b^2) - 4 + 8ab < 0,$$
$$4a^2 + 4b^2 < 4,$$
$$a^2 + b^2 < 1.$$

Thus S is the unit circle (without boundary) and the area is π.

23. (A) We claim that the ratio of consecutive radii is constant, that is, the sequence of radii form a geometric progression. It follows that this ratio is $\sqrt[4]{18/8}$ and that the middle radius is $8\sqrt{18/8} = 12$.

The claim should be intuitive — the picture for any two consecutive circles looks the same except for a change of scale — but here is a proof. In the figure below, three consecutive circles are shown. Their centers P, Q and R are collinear, on the line bisecting the angle formed by the two tangents. Let A, B and C be the points of tangency, and let PS, QT be segments parallel to the upper tangent as shown. Then $\triangle PQS \sim \triangle QRT$. If we let x, y, z be the radii from smallest to largest, then $QS = y-x$, $RT = z-y$. Thus

$$\frac{QS}{PQ} = \frac{RT}{QR} \quad \text{becomes} \quad \frac{y-x}{x+y} = \frac{z-y}{y+z},$$

which simplifies to

$$y^2 = xz, \quad \text{or} \quad \frac{y}{x} = \frac{z}{y}.$$

This last equation says that the ratio of consecutive radii is constant, as claimed.

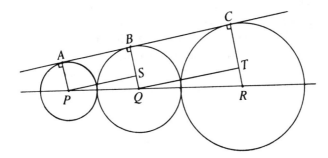

24. (E) We will show that the ratio of perimeter to area in a triangle can be changed at will by replacing the triangle by a similar one. Thus in each class of similar right triangles, there is one with perimeter (in whatever length units, say cm) equal to area (in whatever area units), that is, with the ratio being 1. Since there are infinitely many non-similar right triangles, thus there are infinitely many non-similar (hence non-congruent) right triangles with perimeter = area.

Let a, b, c be the lengths (in cm) of the two sides and the hypotenuse respectively of an arbitrary right triangle. The ratio of perimeter (in cm) to the area (in cm^2) is

$$r = \frac{2(a+b+c)}{ab}.$$

Now consider the similar triangle with sides ka, kb, kc. The ratio is now

$$\frac{2(ka+kb+kc)}{(ka)(kb)} = \frac{1}{k}\frac{2(a+b+c)}{ab} = \frac{r}{k}.$$

Thus, choose $k = r$ and in the new triangle perimeter = area.

Alternative solution sketch. If a and b are the lengths of the legs of a right triangle, then the area and perimeter are numerically equal if and only if

$$\tfrac{1}{2}ab = a + b + \sqrt{a^2 + b^2}.$$

This is one equation in two (nonnegative) real variables. Generally such an equation has infinitely many solutions. Also, one expects different solutions (a, b) to result in non-congruent triangles. Therefore, one should already believe that (E) is correct. A complete analysis (subtle, but left to the reader!) shows that this belief is correct, even though solving the above equation (by rearranging and squaring to remove the radical) introduces extraneous roots, and even though some distinct pairs of valid solutions represent congruent triangles, e.g., $(6, 8)$ and $(8, 6)$.

25. (B) $12 = 60/5 = 60/60^b = 60^{1-b}$. So

$$12^{[(1-a-b)/2(1-b)]} = [60^{1-b}]^{(1-a-b)/2(1-b)}$$

$$= 60^{(1-a-b)/2} = \sqrt{\frac{60}{60^a 60^b}}$$

$$= \sqrt{\frac{60}{3 \cdot 5}} = 2.$$

26. (D) Let $P(E)$ be the probability that event E occurs. By the Inclusion-Exclusion Principle,

$$P(A \cup B) = P(A) + P(B) - P(A \cap B).$$

So

$$p = P(A \cap B) = P(A) + P(B) - P(A \cup B)$$
$$= \frac{3}{4} + \frac{2}{3} - P(A \cup B).$$

At the most, $P(A \cup B) = 1$; at the least,

$$P(A \cup B) = \max\{P(A), P(B)\} = \frac{3}{4}.$$

So

$$\frac{3}{4} + \frac{2}{3} - 1 \le p \le \frac{3}{4} + \frac{2}{3} - \frac{3}{4},$$

which is (D).

Note. Only if A and B are independent is $p = 1/2$.

Remark. Many people assume that if event A has probability p and event B hs probability q, then the probability of both A and B is pq; that is, they assume that events are automatically independent. This problem was devised to make students confront such beliefs, and to make them see just how *d*ependent different events can be.

27. (E) We first construct the figure shown, which results from bisecting the sphere-shadow configuration with a vertical plane. Since the meter stick casts a shadow of $2\,\mathrm{m}$, $AB = 5$ and hence $AD = 5\sqrt{5}$. Now $\triangle CEA$ is similar to $\triangle DBA$. Thus $CE/AC = BD/AD$ or

$$\frac{r}{5 - r} = \frac{10}{5\sqrt{5}} = \frac{2}{\sqrt{5}}.$$

So $\sqrt{5}\,r = 10 - 2r$ or

$$r = \frac{10}{2 + \sqrt{5}} = 10\sqrt{5} - 20.$$

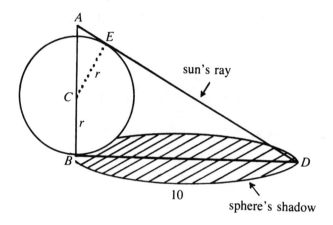

sun's ray

10

sphere's shadow

28. (C)   Draw in line DE. Since

$$\text{Area } ABE = \text{Area } ADE + \text{Area } DBE,$$

and

$$\text{Area } DBEF = \text{Area } FDE + \text{Area } DBE,$$

it follows that Area $ADE =$ Area FDE. Since these two triangles have a common base DE, the altitudes on this base must be equal. That is, A and F are equidistant from line DE, so $AF \parallel DE$. Then, using similar triangles ABC and DBE,

$$\frac{EB}{CB} = \frac{DB}{AB} = \frac{3}{5}.$$

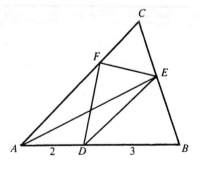

Thus

Area $ABE = \frac{3}{5}$Area $ABC = 6$.

Remark. So much is undetermined in this problem – the shape of the triangle and the posiitions of E and F on their sides – that one might well think that there are many configurations in which ABE and $DBEF$ have the same area and so (E) should be correct. There are indeed many different configurations. For fixed ABC, the solution above shows that if E is in the right place, then F can be *anywhere* on CA. However, the *area* is uniquely determined.

29. (C)   Let the square be placed in the coordinate plane as in the first figure: D is placed at the origin so that algebraic expressions for the distance from it will be easy to interpret. Then $u^2 + v^2 = w^2$ becomes

$$\left[(x-1)^2 + y^2\right] + \left[(x-1)^2 + (y-1)^2\right] = x^2 + (y-1)^2,$$

which simplifies to $x^2 - 4x + 2 + y^2 = 0$, that is, $(x-2)^2 + y^2 = 2$. Thus the locus of p is a circle with center $(2,0)$ and radius $\sqrt{2}$. From the second figure it is clear that the farthest point on this circle from D is $E = (2 + \sqrt{2}, 0)$.

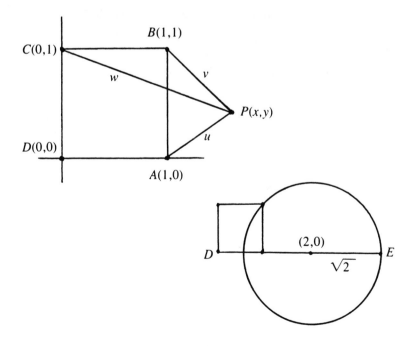

30. (C)   In $\triangle ACP$ and $\triangle BCP$ we have (in the order given) the condition angle-side-side. Since these triangles are not congruent ($\angle CPA \neq \angle CPB$), we must have that $\angle CPA$ and $\angle CPB$ are supplementary. From $\triangle ACP$ we compute

$$\angle CPA = 180° - 10° - (180° - 40°) = 30°.$$

Thus $\angle CPB = 150°$ and

$$\overset{\frown}{BN} = \angle PCB = 180° - 10° - 150° = 20°.$$

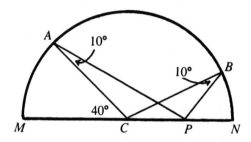

Alternative Solution.   Again $\angle CPA = 30°$. Applying the Law of Sines to $\triangle ACP$ and then $\triangle BCP$, we have

$$\frac{\sin 10°}{CP} = \frac{\sin 30°}{AC} \quad \text{and} \quad \frac{\sin 10°}{CP} = \frac{\sin \angle CPB}{BC}.$$

Thus $\sin \angle CPB = 1/2$ since $AC = BC$. As $\angle CPB \neq \angle CPA$, we must have $\angle CPB = 150°$. Hence $\overset{\frown}{BN} = 20°$.

1984 AHSME Solutions

1. (C)
$$\frac{1000^2}{252^2 - 248^2} = \frac{1000 \cdot 1000}{(252+248)(252-248)}$$
$$= \frac{1000}{500} \cdot \frac{1000}{4} = 500.$$

2. (B)
$$\frac{x - \dfrac{1}{y}}{y - \dfrac{1}{x}} = \frac{\dfrac{xy-1}{y}}{\dfrac{xy-1}{x}} = \frac{xy-1}{y} \cdot \frac{x}{xy-1} = \frac{x}{y}.$$

3. (C)   The number n must be a product of two or more primes greater than 10, not necessarily distinct. The smallest such prime is 11, so the smallest product is $11 \cdot 11 = 121$.

4. (B)   Draw in the perpendicular bisector of chord EF, as shown in the adjoining figure. This bisector must go through the center O of the circle. Since PO is perpendicular to chord BC as well, it bisects BC. Since $PQ \parallel AD$, we have

$$DQ = AP = 4 + \frac{5}{2} = \frac{13}{2}.$$

Thus

$$EF = 2\,EQ = 2(DQ - DE) = 7.$$

OR

Drop perpendiculars BR and CS as shown in the second figure, yielding rectangles $BCRS$ and $ABRD$. Then $ER = 1$, and by symmetry, $SF = 1$. Also, $RS = BC = 5$, so $EF = 1+5+1 = 7$.

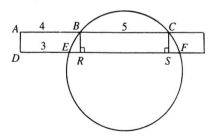

5. (D) Taking the one-hundredth roots,

$$n^{200} < 5^{300} \iff n^2 < 5^3 = 125.$$

The largest perfect square less than 125 is $121 = 11^2$.

6. (B) We have $b = 3g$ and $g = 9t$ (not $3b = g$ or $9g = t$!) so that $g = b/3$ and $t = g/9 = b/27$. Thus the desired sum is (B).

7. (C) Dave travels $(90)(75)$ cm each minute, so the entire trip is $(16)(90)(75)$ cm long. Jack travels $(100)(60)$ cm each minute, so it takes him $(16)(90)(75)/(100)(60)$ or 18 minutes to get to school.

8. (D) Drop perpendiculars from A and B to DC, intersecting DC at F and E, respectively. $\triangle BEC$ is an isosceles right triangle, so $BE = EC = 3$. Since $ABEF$ is a rectangle, $FE = 5$ and $AF = 3$. $\triangle AFD$ is a 30-60-90 triangle, so $DF = AF/\sqrt{3} = \sqrt{3}$. So $DC = 8 + \sqrt{3}$.

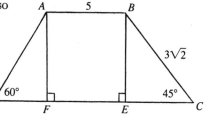

9. (D) $4^{16}5^{25} = 2^{32}5^{25} = 2^7 10^{25} = 128 \cdot 10^{25}$. Thus the usual form is 128 followed by 25 zeros, for a total of 28 digits.

10. (B) Two given points, $1 + 2i$ and $-1 - 2i$, are symmetric with respect to the origin, and the distances from the origin to all three points are equal. Therefore the origin is the center of the square, and the fourth vertex must be symmetric to $-2 + i$ around it. Thus the fourth vertex is $2 - i$.

OR

Plotting the given three points makes it graphically clear where the fourth point must go.

11. (A)   At each stage the displayed entry can be thought of as a power of x. Each reciprocation reverses the sign of the exponent of x. Each squaring doubles the exponent. Thus each pair of squaring and reciprocation multiplies the exponent by -1. Since the initial exponent is 1, the final exponent is $(-2)^n$.

12. (B)   From the definition of the sequence,

$$a_2 - a_1 = 2 \cdot 1$$
$$a_3 - a_2 = 2 \cdot 2$$
$$. \quad . \quad . \quad .$$
$$a_{100} - a_{99} = 2 \cdot 99.$$

Adding, one obtains

$$a_{100} - a_1 = 2(1+2+\cdots+99) = 99 \cdot 100 = 9900,$$
$$a_{100} = 9902.$$

13. (A)   Note that

$$\left(\sqrt{2}+\sqrt{3}+\sqrt{5}\right)\left(\sqrt{2}+\sqrt{3}-\sqrt{5}\right) = \left(\sqrt{2}+\sqrt{3}\right)^2 - 5$$
$$= (5+2\sqrt{6}) - 5 = 2\sqrt{6}.$$

Thus

$$\frac{2\sqrt{6}}{\sqrt{2}+\sqrt{3}+\sqrt{5}} = \frac{2\sqrt{6}}{\sqrt{2}+\sqrt{3}+\sqrt{5}} \cdot \frac{\sqrt{2}+\sqrt{3}-\sqrt{5}}{\sqrt{2}+\sqrt{3}-\sqrt{5}}$$
$$= \sqrt{2} + \sqrt{3} - \sqrt{5}.$$

Remark. In general,

$$\frac{2\sqrt{mn}}{\sqrt{m}+\sqrt{n}+\sqrt{m+n}} = \sqrt{m} + \sqrt{n} - \sqrt{m+n}.$$

14. (A)   Given $x^{\log_{10} x} = 10$,   take logs base 10 of both sides:

$$(\log_{10} x)(\log_{10} x) = 1,$$
$$\log_{10} x = \pm 1,$$
$$x = 10, 10^{-1}.$$

Neither solution is extraneous. Thus the product of the roots is 1.

15. (A)   The following statements are equivalent:

$$\sin 2x \sin 3x = \cos 2x \cos 3x,$$
$$\cos 2x \cos 3x - \sin 2x \sin 3x = 0,$$
$$\cos(2x+3x) = 0,$$
$$5x = 90° + 180°k, \quad k = 0, \pm 1, \pm 2, \ldots,$$
$$x = 18° + 36°k, \quad k = 0, \pm 1, \pm 2, \ldots.$$

The only correct value listed among the answers is 18°.

OR

By inspection of the original equation, it is sufficient that $\sin 2x = \cos 3x$ and $\sin 3x = \cos 2x$, which are both true if $2x$ and $3x$ are complementary. Thus $2x+3x = 90°$, i.e., $x = 18°$, is a correct value.

16. (E)   If $r_1 = 2+a$ is a root, then so is $r_2 = 2-a$. That is, all the distinct roots come in pairs symmetric around $x = 2$ (except for the self-symmetric value $r = 2$, if it is a root). Each pair has sum 4. Since there are exactly four distinct roots, 2 is not a root, there are exactly two pairs, and the sum of all roots is 8.

Note. See the related AIME problem #12 from 1984.

17. (C)   A leg of a right triangle is the geometric mean of the hypotenuse and the projection of the leg on the hypotenuse. Setting $AH = x$, it follows that

$$225 = x(x+16),$$
$$x^2 + 16x - 225 = 0,$$
$$(x+25)(x-9) = 0,$$
$$x = 9.$$

Thus $AB = 25$, $CH = \sqrt{15^2-9^2} = 12$, and the area of $\triangle ABC$ is $\frac{1}{2} \cdot 25 \cdot 12 = 150$.

18. (E) From a picture we can determine that there must be four such points, with different x coordinates. To be equidistant from two nonparallel lines L and K, a point must be on *either* of the two lines which bisect a pair of opposite angles at the intersection of L and K. In particular, to be equidistant from the x and y axes, (x,y) must

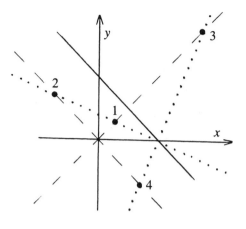

be on either $y = x$ or $y = -x$ (shown dashed in the adjoining figure); to be equidistant from the x-axis and $x+y = 2$, (x,y) must be on one of the lines through $(2,0)$ shown dotted in the figure. There are four points simultaneously on one of the dashed and one of the dotted lines, as indicated. One can show that the x coordinates of the four points, in the order labeled, are $2 - \sqrt{2}$, $-\sqrt{2}$, $2 + \sqrt{2}$ and $\sqrt{2}$.

OR

The line and the axes determine a triangle. This triangle has an inscribed circle and three escribed circles. The centers of these circles, and no other points, satisfy the equal-distance condition. Hence there are four points. It is easy to see by a sketch that their x-coordinates are not all the same.

19. (D) The sum is odd if and only if an odd number of odd numbered balls is chosen. Of the $\binom{11}{6} = 462$ possible unordered set of 6, $\binom{6}{1}\binom{5}{5} = 6$ have 1 odd numbered ball, $\binom{6}{3}\binom{5}{3} = 200$ have 3, and $\binom{6}{5}\binom{5}{1} = 30$ have 5. So the probability is

$$\frac{6+200+30}{462} = \frac{118}{231}.$$

20. (C) The equation $\left| x - |2x+1| \right| = 3$ is satisfied by precisely those x for which the vertical distance between the graph of $y = x$ and the graph of $y = |2x+1|$ is 3. These two graphs are sketched in the adjoining figure; the latter graph is V-shaped with vertex $(-\frac{1}{2}, 0)$ and sides of slope ± 2. Thus there are exactly two x values which satisfy the original equation, because the vertical distance between the graphs increases steadily as one moves right or left from $x = -\frac{1}{2}$, and the vertical distance for that x is $\frac{1}{2}$.

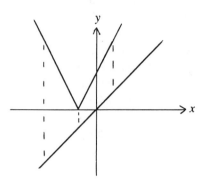

OR

Either $x - |2x+1| = -3$ or $x - |2x+1| = 3$, yielding $|2x+1| = x + 3$ and $|2x+1| = x - 3$, respectively. In the first case one must have $x \geq -3$, while in the second case, $x \geq 3$ must hold. In the first case, $2x + 1$ must equal $x+3$ or $-x-3$, yielding $x = 2$ and $x = -4/3$, which are solutions. In the second case, $2x + 1$ must equal $x-3$ or $-x+3$, and hence $x = -4$ and $x = 2/3$; however, these values violate the condition $x \geq 3$ and therefore do not satisfy the original equation.

Query. How many solutions are there if x is allowed to be complex, and $|w|$ for a complex number is defined as in Problem 30 of this exam? (Problem 14 did not explicitly restrict x to be real, and one student wrote to us about this, but fortunately for the Contest Committee, the correct answer when x may be complex is not among the choices listed.)

21. (C) From the second equation, $c(a+b) = 23$ and 23 is prime. Consequently, the two factors must be 1 and 23. Since a and b are positive integers, $a + b > 1$. Hence one must have $c = 1$ and $a + b = 23$. Upon substituting 1 for c and $23 - a$ for b into the first equation, it becomes a quadratic, $a^2 - 22a + 21 = 0$, with solutions $a = 1$ and $a = 21$. Both of these, and the corresponding values of b (22 and 2), satisfy both equations. Thus the solutions are (1,22,1) and (21,2,1).

22. (B)   The vertex of any parabola $y = ax^2 + tx + c$ is on its axis of symmetry $x = -t/2a$. Thus $x_t = -t/2a$ and

$$y_t = a\left(\frac{-t}{2a}\right)^2 + t\left(\frac{-t}{2a}\right) + c = c - \frac{t^2}{4a} = c - ax_t^2.$$

Thus every point (x_t, y_t) is on the parabola $y = -ax^2 + c$. Conversely, each point (x, y) on this parabola is of the form (x_t, y_t): just set $t = -2ax$.

23. (D)   Two trigonometric identities for expressing sums as products are:

$$\sin x + \sin y = 2 \sin \frac{x+y}{2} \cos \frac{x-y}{2},$$
$$\cos x + \cos y = 2 \cos \frac{x+y}{2} \cos \frac{x-y}{2}.$$

Thus

$$\frac{\sin 10° + \sin 20°}{\cos 10° + \cos 20°} = \frac{\sin 15°}{\cos 15°} = \tan 15°.$$

24. (E)   That the roots of $x^2 + ax + 2b = 0$ are both real implies that the discriminant $a^2 - 8b$ is nonnegative, that is, $a^2 \geq 8b$. Likewise, that $x^2 + 2bx + a = 0$ has real roots implies $4b^2 \geq 4a$ or $b^2 \geq a$. Thus $a^4 \geq 64b^2 \geq 64a$. Since $a > 0$, one obtains $a^3 \geq 64$ and so $a \geq 4$ and $b \geq 2$. Conversely, $a = 4$, $b = 2$ does result in real roots for both (identical) equations. Thus the minimum value of $a + b$ is 6.

25. (D)   With x, y, z, d as in the adjoining figure, we have

(1) $2xy + 2xz + 2yz = 22$,

(2) $4x + 4y + 4z = 24$,

(3) $d^2 = x^2 + y^2 + z^2$.

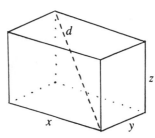

From (2) we obtain

$(x + y + + z)^2 = 6^2$,   or

(4) $x^2 + y^2 + z^2 + 2xy + 2xz + 2yz = 36$.

Substituting (1) and (3) into (4) gives $d^2 + 22 = 36$, or $d = \sqrt{14}$.

Note. The dimensions of the solid are not uniquely determined, but d is!

26. (B) Draw in MC as in the adjoining figure. Then $\triangle DMC$ and

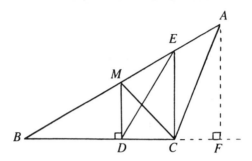

$\triangle DME$ have the same area, as they have the same base MD and equal altitudes on that base. Thus

$$\text{Area } \triangle BMC = \text{Area } \triangle BED.$$

Moreover, as shown below,

$$\text{Area } \triangle BMC = \tfrac{1}{2}\text{Area } \triangle BAC.$$

Thus

$$\text{Area } \triangle BED = \tfrac{1}{2}\text{Area } \triangle BAC = 12.$$

As for the claim in the second display, since M is the midpoint of BA, altitude MD of $\triangle BMC$ is one half altitude AF of $\triangle BAC$. Since the triangles have the same base BC, the claim follows.

Alternatively, one proves the claim by noting that

$$\text{Area } \triangle BMC = \tfrac{1}{2}(BM)(BC)\sin B = \tfrac{1}{2}\left[\tfrac{1}{2}(BA)(BC)\sin B\right]$$
$$= \tfrac{1}{2}\text{Area } \triangle BAC.$$

27. (C) Let $AC = x$ and $\angle DCF = \theta$. Then $\angle CBD = \theta$, and $\angle ADB = 2\theta$ by the Exterior Angle Theorem. Thus $\cos\theta = 1/x$ and $\cos 2\theta = x-1$. Therefore,

$$2\left(\frac{1}{x}\right)^2 - 1 = x - 1,$$
$$2 - x^2 = x^3 - x^2,$$
$$x = \sqrt[3]{2}.$$

OR

Drop $DG \perp BC$. Let $AC = x$, $GC = y$. Note that $BC = 2y$, for $\triangle BDC$ is isosceles. Since $\triangle DCG \sim \triangle ACF \sim \triangle BCA$, we obtain $\dfrac{1}{y} = \dfrac{x}{1} = \dfrac{2y}{x}$.

Thus $y = 1/x$ and $y = x^2/2$, implying $x^3 = 2$, or $x = \sqrt[3]{2}$.

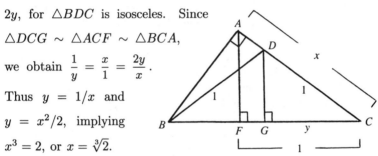

Query. How can $\sqrt[3]{2}$ be the answer to this problem when doubling the cube (i.e., constructing $\sqrt[3]{2}$) is known to be impossible?

Answer. It is true that constructing $\sqrt[3]{2}$ is impossible, if one means constructing a segment of that length, starting with a segment of length 1 and using compass and straightedge *only*. But just because such a construction is impossible, that doesn't mean a segment can't have length $\sqrt[3]{2}$. In fact, if one is allowed to use a *ruler* instead of a straightedge (a ruler has markings, a straightedge doesn't), then one *can* construct $\sqrt[3]{2}$. See Heinrich Dorrie, "100 Great Problems of Elementary Mathematics," Section 35 (Dover Pub., New York, 1965). Even if mankind knew no way to construct $\sqrt[3]{2}$, that still wouldn't mean that it doesn't exist.

Another good source of information on constructibility is Ivan Niven, "Numbers: Rational and Irrational," New Mathematical Library, Vol. 1, Mathematical Association of America, Washington, DC.

28. (C) Note that the prime factorization of 1984 is $2^6 \cdot 31$, that $x < 1984$ and that

$$y = (\sqrt{1984} - \sqrt{x})^2 = 1984 + x - 2\sqrt{1984x}\,.$$

It follows that y is an integer if and only if $1984x$ is a perfect square, that is, if and only if x is of the form $31t^2$. Since x is less than 1984, we have $1 \le t \le 7$, yielding the pairs

$$(31, 1519), \quad (124, 1116) \quad \text{and} \quad (279, 775)$$

for (x, y), corresponding to $t = 1, 2, 3$. Since $y \le x$ for $t > 3$, these are the only solutions.

29. (A) Let $P = (x, y)$, $A = (0, 0)$, $C = (3, 3)$ and B be any point on the positive x-axis. The locus of P is the circle with center C and radius $\sqrt{6}$, and y/x is the slope of segment AP. Clearly this slope is greatest when AP is tangent to the circle on the left side, as in the adjoining figure (Note: $\sqrt{6} < 3$). Let $\alpha = \angle CAP$. Since $\angle BAC = 45°$, the answer is

$$\tan(\alpha + 45°) = \frac{\tan\alpha + 1}{1 - \tan\alpha}\,.$$

Since $\angle APC = 90°$, $\tan\alpha = PC/PA$. By the Pythagorean Theorem

$$PA = \sqrt{(AC)^2 - (PC)^2} = 2\sqrt{3}\,.$$

Thus $\tan\alpha = 1/\sqrt{2}$ and the answer is $3 + 2\sqrt{2}$.

OR

The maximum value of y/x is the maximum value of the slope of a line which contains the origin and which intersects the circle $(x-3)^2 + (y-3)^2 = 6$. Clearly the line of maximum slope, m, is tangent to the circle and is the steeper of the two lines through the origin and tangent to the circle. In short, it is the line containing AP in the previous figure. The following statements are equivalent:

1) $y = mx$ is tangent to the circle;

2) the system of equations

$$y = mx,$$
$$(x-3)^2 + (y-3)^2 = 6,$$

has only one solution (x, y);

3) the quadratic equation $(x-3)^2 + (mx-3)^2 = 6$, or

$$(m^2+1)x^2 - 6(m+1)x + 12 = 0,$$

has a double root;

4) the discriminant is zero, i.e.,

$$36(m+1)^2 - 48(m^2+1) = 0.$$

Solving the last equation for its larger root, one obtains $m = 3 + 2\sqrt{2}$.

Note. This method is more general than the previous one. The circle may be replaced by any conic and the method will still work. Also, this problem may of course be attacked by calculus, but unless one is very careful setting the problem up, the algebra gets extremely involved.

30. (B) We derive a more general result: if n is an integer > 1 and

$$w = \cos \frac{2\pi}{n} + i \sin \frac{2\pi}{n},$$

then

$$\left| w + 2w^2 + \cdots + nw^n \right|^{-1} = \frac{2}{n} \sin \frac{\pi}{n}.$$

To prove this, let

$$S = w + 2w^2 + \cdots + nw^n.$$

Multiplying both sides of this equation by w and subtracting Sw from S, we get

$$S(1-w) = w + w^2 \cdots + w^n - nw^{n+1}.$$

Now, $w \neq 1$ since $n > 1$. Thus we may use the formula for summing a geometric series to obtain

$$S(1-w) = \frac{w^{n+1} - w}{w - 1} - nw^{n+1}.$$

Since $w^n = 1$ (by DeMoivre's formula), this reduces further to $S(1-w) = -nw$. Thus

$$\frac{1}{S} = \frac{w-1}{nw}, \qquad \frac{1}{|S|} = \frac{|w-1|}{n}.$$

Finally, $|w-1|$ is the length of the side of the regular n-gon inscribed in the unit circle, since 1 and w are consecutive vertices. It is well known that this side length is $2\sin\frac{\pi}{n}$ (in the isosceles triangle with vertices 0, 1, w, drop an altitude from 0).

1985 AHSME Solutions

1. (A) $4x + 1 = 2(2x + 1) - 1 = 2 \cdot 8 - 1 = 15$. Or, solve $2x + 1 = 8$ to obtain $x = 7/2$. Thus $4x + 1 = 4(7/2) + 1 = 15$.

2. (E) The shaded sector has central angle $360° - 60°$, or $\frac{5}{6}$ of $360°$. Thus the perimeter consists of $\frac{5}{6}$ of the circumference of a circle of radius 1, plus 2 radii of the circle. Thus the length is

$$\frac{5}{6}(2\pi r) + 2r = \frac{5}{3}\pi + 2.$$

3. (D) Since $\triangle ABC$ is a right triangle, $AB = 13$. Also, $BN = BC = 5$ and $AM = AC = 12$. Thus

$$BM = AB - AM = 1,$$
$$MN = BN - BM = 4.$$

4. (C) If the bag contains p pennies, d dimes and q quarters, then $d = 2p$ and $q = 3d$. Thus $q = 6p$. If A is the worth of the coins in cents, then

$$A = p + 10(2p) + 25(6p) = 171p.$$

Of the answers given, only (C) is a multiple of 171.

5. (E) $$\frac{1}{2} + \frac{1}{4} + \frac{1}{6} + \frac{1}{8} + \frac{1}{10} + \frac{1}{12}$$
$$= \frac{60}{120} + \frac{30}{120} + \frac{20}{120} + \frac{15}{120} + \frac{12}{120} + \frac{10}{120}.$$

Since $60 + 30 + 20 + 10 = 120$, it is clear that one *may* remove

$$\frac{15}{120} = \frac{1}{8} \quad \text{and} \quad \frac{12}{120} = \frac{1}{10}.$$

Furthermore, one sees easily from the right hand side above that no other terms sum to $\frac{27}{120}$. Thus one *must* remove $\frac{1}{8}$ and $\frac{1}{10}$.

6. (B)   Since the selection is random, the *number* of boys is 2/3 the number of girls. Thus there are 2 boys for every 3 girls, i.e., 2 boys for every 5 students. Alternately, let b and g be the number of boys and girls. Then $b = \frac{2}{3}g$ and so

$$\frac{b}{b+g} = \frac{(2/3)g}{(5/3)g} = \frac{2}{5}.$$

7. (E)   Going from right to left, we first add c to d, obtaining $c+d$. Then we subtract this result from b, obtaining $b-c-d$. Finally, we divide a by this last expression, obtaining (E).

8. (E)   The solution of $ax+b = 0$ is $-b/a$. The solution of $a'x+b' = 0$ is $-b'/a'$. Thus the question becomes: Which one of the five inequalities in the answers is equivalent to

$$\frac{-b}{a} < \frac{-b'}{a'} ? \qquad\qquad (*)$$

Multiplying by -1, one obtains

$$\frac{b}{a} > \frac{b'}{a'},$$

which is (E). Conversely, multiplying (E) by -1 gives $(*)$, so they are equivalent.

Note. Multiplying $(*)$ by $-aa'$ results in (B) if, but only if, $aa' > 0$. Thus $(*)$ and (B) are not equivalent.

9. (B)   Observe that the first column consists of all numbers with value $16n-1$ for $n = 1, 2, \ldots$, and that the first number in the next row beneath each value of $16n-1$ is $16n+1$ for the same n and appears in the second column. Since $1985 = 16 \cdot 124 + 1$, 1985 appears in the second column.

OR

Observe that all the odd integers between 0 and 8 appear in the first row, all between 8 and 16 in the second, and in general, all between $8(n-1)$ and $8n$ in the nth. Since $1985 = 248 \cdot 8 + 1$, it follows that 1985 is the smallest odd integer in the 249th row. Now observe that the entries in odd-numbered rows increase to the right and start in the second column.

10. (E)   By definition, the graph of $y = f(x)$ is the set of points (x, y) in the rectilinear coordinate plane for which $y = f(x)$. Thus the graph of $y = \sin x$ is the usual periodic wave along the x-axis. Consider a circle tangent to the x-axis at the origin. If the radius is *very* large, the circle stays close to the x-axis for a long time and thus intersects the sine curve many times. By making the radius arbitrarily large, the number of intersection points may be made arbitrarily large.

11. (B)   There are 2 ways to order the vowels. There would be $5! = 120$ orderings of the consonants if they were all distinct. Since there are two T's, there are $120/2 = 60$ consonant orderings. Thus there are $2 \times 60 = 120$ reorderings of **CONTEST** with the vowels first.

12. (D)   If $n = pq^2r^4$ is a divisor of cube c, then c must have p, q and r as primes in its factorization. Moreover, the exponents of p, q and r in the factorization must be multiples of 3 and they must be at least as great as 1, 2 and 4, respectively (1, 2, 4 being the exponents of p, q, r in n). Thus $p^3q^3r^6 = (pqr^2)^3$ is the smallest such cube.

13. (E)   We give three methods.

Method 1. Using the dotted horizontal and vertical lines shown as part of the figure, divide $PQRS$ into 4 right triangles, I, II, III, IV. Compute the area of each from the lengths of the legs and add.

Method 2.   Surround $PQRS$ by rectangle $ABCD$. Note that for each right triangle in Method 1, there is a second congruent triangle in $ABCD$. Thus

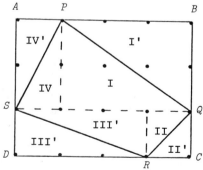

Area $PQRS = \frac{1}{2}$Area $ABCD = \frac{1}{2}(3 \times 4) = 6.$

Method 3. Let P be any polygon in the plane which has all its vertices at lattice points (points with both coordinates integers). P can have any number of sides, and need not be convex, but different sides cannot cross each other. Then

$$\text{Area } = I + \tfrac{1}{2}B - 1,$$

where I is the number of lattice points strictly inside P and B is the number on the boundary. Thus

$$\text{Area } PQRS = 5 + \tfrac{1}{2}(4) - 1 = 6.$$

This surprising formula is called *Pick's Theorem*. A proof may be found in Duane DeTemple & Jack Robertson, *The Equivalence of Euler's and Pick's Theorems*, The Math. Teacher, March 1974, pp. 222-226. Also see Ross Honsberger, "Ingenuity in Mathematics," New Mathematical Library, Vol. 23, pp. 27-31.

14. (C) For any convex n-gon, the sum of the interior angles is $(n-2)180°$. If an n-gon has exactly 3 obtuse interior angles, then the remaining $n-3$ angles have measure at most 90° each and the obtuse angles have measure less than $3 \cdot 180°$ together. Thus

$$(n-2)180 < (n-3)90 + 3 \cdot 180,$$
$$2(n-2) < n-3 + 6,$$
$$n < 7.$$

To show $n = 6$ is possible, "distort" an equilateral triangle as shown.

Alternative Solution (from John Reutershan, teacher, Bronx High School of Science, NY). There are exactly 3 obtuse interior angles if and only if there are exactly 3 acute exterior angles. In any convex polygon, the sum of all the exterior angles is 360°. So if there are exactly 3 acute exterior angles, there can be at most 3 nonacute exterior angles (because less than 360° are left and each nonacute angle has at least 90°). Thus $n = 6$ sides is an upper bound; the figure for the first solution shows that 6 is possible.

15. (E) Raise both sides of $a^b = b^a$ to the power $1/a$ and substitute $b = 9a$ to obtain $a^9 = 9a$. Since $a \neq 0$, $a^8 = 9$; so $a = 9^{1/8} = (3^2)^{1/8} = 3^{1/4}$.

Note. For any positive $k \neq 1$, the equations $a^b = b^a$, $b = ka$ have the unique positive simultaneous solution

$$(a, b) = \left(k^{1/(k-1)}, k^{k/(k-1)}\right).$$

16. (B) We show that for any angles A and B for which the tangent function is defined and $A+B = 45°$, $(1 + \tan A)(1 + \tan B) = 2$. By the addition law for tangents,

$$1 = \tan 45° = \tan(A+B) = \frac{\tan A + \tan B}{1 - \tan A \tan B},$$

$$1 - \tan A \tan B = \tan A + \tan B,$$

$$1 = \tan A + \tan B + \tan A \tan B.$$

Thus

$$(1 + \tan A)(1 + \tan B) = 1 + \tan A + \tan B + \tan A \tan B$$
$$= 1 + 1 = 2.$$

17. (B) Triangles I and II in the figure are similar. Thus $\frac{1}{b} = \frac{b}{2}$ and $b = \sqrt{2}$. The area of the rectangle is twice the sum of the areas of I and II, so the answer is $3b = 3\sqrt{2} \approx 3(1.414) \approx 4.2$.

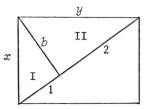

18. (D) Jane gets twice as many marbles as George, so the total number of marbles taken must be divisible by 3; that is, the difference between 140 and the number of chipped marbles must be divisible by 3. Of the six given numbers only 23 has this property. Therefore, there are 23 chipped marbles, leaving $19 + 25 + 34 = 78$ for Jane and $18 + 21 = 39$ for George.

19. (A)   Rearrange the terms and complete the square for the second equation to obtain

$$(y-2)^2 - x^2 = 1,$$

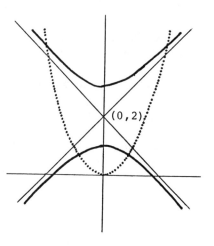

whose graph is a hyperbola with center $(0,2)$, vertices at $(0,1)$ and $(0,3)$, and asymptotes $y = 2 \pm x$. The graph of $y = Ax^2$ is a parabola opening upward with its vertex at the origin for all positive values of A. Therefore there are two distinct points of intersection between the parabola and the lower branch of the hyperbola. The equation of the upper branch of the hyperbola is $y = 2 + \sqrt{x^2 + 1}$. For any given positive A, if the absolute value of x is large enough, then

$$Ax^2 > 2 + \sqrt{x^2 + 1} \approx 2 + x.$$

Since $Ax^2 < 2 + \sqrt{x^2 + 1}$ when $x = 0$, the parabola must also intersect the upper branch of the hyperbola in two distinct points for any given positive value of A.

OR

We show that the system of equations

$$(1) \quad y = Ax^2, \qquad (2) \quad y^2 + 3 = x^2 + 4y,$$

has exactly four distinct solutions for any $A > 0$. First, multiply (2) by A, eliminate Ax^2 using (1), and simplify, obtaining

$$(3) \quad Ay^2 - (4A+1)y + 3A = 0.$$

Since all these steps are reversible, the system (1) and (3) has the same solution pairs as the system (1) and (2). Furthermore, (1) and (3) has two distinct solution pairs (x, y) for each positive solution y to (3) alone, namely, $(\sqrt{y/A}, y)$ and $(-\sqrt{y/A}, y)$. So it suffices to show that (3) has exactly two solutions, both positive. By the quadratic formula,

$$(4) \quad y = \frac{4A+1 \pm \sqrt{(4A+1)^2 - 12A^2}}{2A}.$$

Since the discriminant in (4) is positive for $A > 0$, there are two real roots. Since the constant term of (3) is positive and the coefficient of y is negative, both these roots must be positive.

20. (D) A smaller cube will have no face painted if it comes from the interior of the original cube, that is, if it is part of the cube of side length $n - 2$ obtained by stripping away a one-unit layer from each face of the original. So there are $(n-2)^3$ unpainted smaller cubes. A smaller cube has one painted face if it comes from one of the faces of the original cube, but was not on the edge of the face. Thus there are $6(n-2)^2$ such smaller cubes. Therefore, $(n-2)^3 = 6(n-2)^2$. Since $n > 2$, we can divide out the $(n-2)^2$, leaving $n - 2 = 6$, or $n = 8$.

21. (C) There are 3 ways a^b can equal 1 when a and b are integers:

   I) $a = 1$;

   II) $a = -1$, b even;

   III) $b = 0$, $a \neq 0$.

In this problem $a = x^2 - x - 1$, $b = x + 2$. We solve each case in turn.

Case I: $$x^2 - x - 1 = 1,$$
$$(x-2)(x+1) = 0,$$
$$x = 2 \text{ or } 1.$$

Case II: $$x^2 - x - 1 = -1 \text{ and } x + 2 \text{ is even}$$
$$x^2 - x = 0,$$
$$x = 0 \text{ or } 1.$$

The choice $x = 0$ is a solution, since then $x + 2$ is even. However, $x = 1$ is not a solution, since then $x + 2$ would be odd.

Case III: $$x + 2 = 0 \text{ and } x^2 - x - 1 \neq 0,$$
$$x = -2.$$

This is a solution, since $(-2)^2 - (-2) - 1 = 5 \neq 0$.

Thus there are 4 solutions in all.

22. (D)   Consider the half circle $ACDO$. Let $2\theta = 60° = \angle ABO = \overset{\frown}{CD}$. Then $\angle CAD = \theta$, since it is an inscribed angle. Draw in CO. Since $\angle COA$ is isosceles, $\angle ACO = \theta$. By the Exterior Angle Theorem applied to $\triangle BOC$ at B, $\angle BOC = 2\theta - \theta = \theta$. So $\angle BOC$ is also isosceles and $BC = 5$.

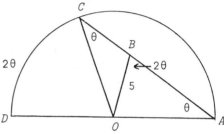

Note. The fact that $2\theta = 60°$ is irrelevant. $BC = 5$ whatever θ is. (What θ's are possible?) However, it may be psychologically easier to solve the problem with $2\theta = 60°$. For instance, $\angle BAO = 30°$ (inscribed angle) so $\angle BOA = 90°$. Thus, $\angle BOC = 180° - 60° - 90° = 30°$. Next, $\angle CBO = 180° - \angle ABO = 120°$. So $\angle BCO = 180° - (30° + 120°) = 30°$. Thus $CB = BO = 5$.

23. (C)   Note that $x^3 = y^3 = 1$, because x and y are the complex roots of $0 = x^3 - 1 = (x-1)(x^2+x+1)$. Alternatively, plot x and y in the complex plane and observe that they have modulus 1 and arguments $2\pi/3$ and $4\pi/3$. Thus $x^9 + y^9 = 1 + 1 \neq -1$. (To show that the other equations are correct is easy if one also notes that $y = x^2$ and $x + y = -1$. For instance, $x^{11} + y^{11} = x^2 + x^{22} = y + x = -1$.)

24. (C)   Let $\Pr\{d_1, d_2, \ldots\}$ be the probability that the digit chosen is one of $d_1, d_2, \ldots$. Note that

$$\Pr\{d\} = \log \frac{d+1}{d}$$

and

$$\Pr\{d, d+1\} = \log \frac{d+1}{d} + \log \frac{d+2}{d+1} = \log \frac{d+2}{d}.$$

We seek a set of digits with probability $2\Pr\{2\}$. Thus

$$2\Pr\{2\} = 2\log\frac{3}{2} = \log\frac{9}{4}$$
$$= \log\frac{5}{4} + \log\frac{6}{5} + \cdots + \log\frac{9}{8}$$
$$= \Pr\{4, 5, 6, 7, 8\}.$$

Note 1. In this solution, we have not used the fact that the logs are base 10. However, this fact is necessary for the problem to make sense; otherwise the union of all possibilities (i.e, picking some digit from 1 to 9) does not have probability 1.

Note 2. If one collects a lot of measurements from nature (say, the lengths of American rivers in miles), the fraction of the time that the first significant (i.e., nonzero) digit is d is approximately $\log_{10}(d+1) - \log_{10} d$. In particular, the distribution is *not* uniform, e.g., 1 is the first digit about 30% of the time. This counterintuitive fact, sometimes called Benford's Law, can be explained if one assumes that the distribution of natural constants is independent of our units of measurement. See two articles by R. Raime: *On the Distribution of First Significant Figures*, Amer. Math. Monthly 76 (1969), pp. 342-48, and *The Peculiar Distribution of First Digits*, Scientific Amer., Dec. 1969, p. 109ff. For further explanations of Benford's Law, see Section 4 of *Some Remarkable Sequences of Integers*, by R. P. Boas, in *Mathematical Plums*, Ross Honsbrger, ed., The Mathematical Association of America, Washington DC, 1979.

25. (B)   Label the edges of the solid a, ar, and ar^2. Then

Volume $= a(ar)(ar^2) = 8$,
   $ar = 2.$

Surface Area
   $= 2a^2r + 2a^2r^2 + 2a^2r^3 = 32$
   $= 2(ar)(a + ar + ar^2)$
   $= 4(a + ar + ar^2).$

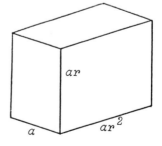

But this last expression is just the sum of the edge lengths.

Note 1. One can also solve for a and r, and thus find that the edges have lengths $3 - \sqrt{5}$, 2 and $3 + \sqrt{5}$.

Note 2. For any rectangular solid with volume 8 and edge lengths in geometric progression, the surface area will equal the sum of the edge lengths.

26. (E) $\frac{n-13}{5n+6}$ is reducible and nonzero iff its reciprocal $\frac{5n+6}{n-13}$ exists and is reducible. By long division, $\frac{5n+6}{n-13} = 5 + \frac{71}{n-13}$. Thus it is necessary and sufficient that $\frac{71}{n-13}$ be reducible. Since 71 is a prime, $n-13$ must be a multiple of 71. So $n-13 = 71$, or $n = 84$, is the smallest solution.

OR

We seek the smallest $n > 0$ for which $n-13$ and $5n+6$ have a common factor and $n-13 \neq 0$. To make it easier to see a common factor, set $m = n-13$; then $5n+6 = 5m+71$. Clearly, m and $5m+71$ have a common factor iff m and 71 do. Since 71 is a prime, m must be one of $\ldots, -71, 0, 71, 142, \ldots$. Thus n must be one of $\ldots, -58, 13, 84, 155, \ldots$. The smallest positive value of n giving a positive fraction is $n = 84$.

Third Solution (from John Reutershan, teacher, Bronx High School of Science, NY). If $\frac{n-13}{5n+6}$ is reducible, then there is some prime p that divides both $n-13$ and $5n+6$. But then p divides any linear combination of these terms; in particular, p divides $(5n+6) - 5(n-13) = 71$. Thus $p = 71$ and $n-13 = 71k$ for some integer k. Since $n > 0$ and $n-13 \neq 0$, n must be at least 84.

27. (C)   We have

$$x_1 = 3^{1/3},$$
$$x_2 = \left(3^{1/3}\right)^{\sqrt[3]{3}} = 3^{\sqrt[3]{3}/3},$$
$$x_3 = \left(3^{\sqrt[3]{3}/3}\right)^{\sqrt[3]{3}} = 3^{\sqrt[3]{9}/3},$$
$$x_4 = \left(3^{\sqrt[3]{9}/3}\right)^{\sqrt[3]{3}} = 3^{\sqrt[3]{27}/3} = 3^{3/3} = 3.$$

One should verify that x_1, x_2 and x_3 are not integers. Here is a sketch of one way to do this. First, prove by induction that the sequence $\{x_n\}$ is increasing. Next, since $1 < x_1 < 2$ and $x_4 = 3$, it suffices to show that neither x_2 nor x_3 is 2. As for x_2, note that $x_1 < 3/2$ (since $x_1{}^3 = 3 < 27/8$). Thus

$$x_2 = x_1^{x_1} < (3/2)^{3/2} = \sqrt{27/8} < 2.$$

To show $x_3 > 2$, show that $x_1 > \sqrt{2}$ and do a similar manipulation.

Note. x_2 and x_3 are integers raised to radicals, e.g.,

$$x_2 = 3^{(\sqrt[3]{1/9})}$$

There is a general theorem which says (as a special case) that such numbers cannot be integers (except when the base is 0 or 1), but the proof is very deep. For an introduction, see I. Niven, "Numbers: Rational and Irrational," New Mathematical Library, Vol. 1, Section 5.4.

28. (B) By the Law of Sines,

$$\frac{27}{\sin A} = \frac{48}{\sin 3A}.$$

Using the identity

$$\sin 3A = 3\sin A - 4\sin^3 A,$$

we have

$$\frac{48}{27} = \frac{16}{9} = \frac{\sin 3A}{\sin A} = 3 - 4\sin^2 A.$$

Solving for $\sin A$ gives $\sin A = \sqrt{11}/6$ and $\cos A = 5/6$. ($\cos A$ cannot be negative since $0 < 3A < 180°$.) Again by the Law of Sines,

$$\frac{b}{\sin(180°-4A)} = \frac{27}{\sin A} \quad \text{or} \quad b = \frac{27\sin 4A}{\sin A}.$$

Since

$$\sin 4A = 2\sin 2A \cos 2A = 4\sin A \cos A \,(\cos^2 A - \sin^2 A),$$

we have

$$b = 27 \cdot 4 \cdot \frac{5}{6}\Big(\frac{25-11}{36}\Big) = 35.$$

OR

Divide $\angle C$ into $\alpha = \angle A$ and 2α as shown in the figure. By the Exterior Angle Theorem, $\angle CDB = 2\alpha$. Thus $DB = CB = 27$ and $AD = 48-27 = 21$. Since $\triangle ADC$ is isosceles also, $CD = 21$. Now apply Stewart's Theorem [see H. S. M. Coxeter &

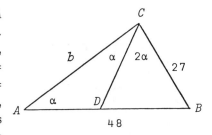

S. L. Greitzer, "Geometry Revisited," New Mathematical Library, Vol. 19], which says that for *any* point D on AB,

$$(AC)^2(BD) + (BC)^2(AD) = (AB)\left[(AD)(BD)+(CD)^2\right].$$

Thus $27b^2 + 27^2 \cdot 21 = 48\left(21 \cdot 27 + 21^2\right)$. Solving gives $b = 35$.

Note. If the vertices of a triangle were allowed to be collinear (i.e., if we allowed "degenerate" triangles with vertex angles of $0°$ and $180°$), then there is another solution: $b = 48 + 27 = 75$, with B on the line segment AC. Thus the correct answer would be (E), not uniquely determined. However, high school texts and the ASHME examination have always assumed that triangles are not degenerate, and no participant wrote to us to argue for answer (E). (The issue was brought to our attention by Prof James Brooks of Villanova University.) Nonetheless, this new solution raises interesting questions. Degenerate triangles (and other degenerate figures) are useful for some purposes, and the Law of Sines is valid for degenerate triangles; so why didn't the first solution above reveal the degenerate answer? Reason: the Law of Sines is correct in the degenerate case only if it is written in inverted form to avoid 0 in the denominator. If we start with $\frac{\sin A}{27} = \frac{\sin 3A}{48}$, it follows that one solution is $\sin A = 0$.

Query. Where is the "flaw" in the second solution above; why doesn't it find the degenerate case?

29. (C) First note that since $9ab$ ends in a zero, the sum of its digits is the same as the sum of the digits in $N = 9ab/10$. Second note that for any integer M represented as a string of k copies of the digit d, $M = \frac{d}{9}(999\ldots 99) = \frac{d}{9}(10^k - 1)$. Thus

$$N = \tfrac{9}{10}\big(\tfrac{8}{9}(10^{1985}-1)\big)\big(\tfrac{5}{9}(10^{1985}-1)\big)$$
$$= \tfrac{4}{9}(10^{2\cdot 1985} - 2\cdot 10^{1985} + 1)$$
$$= \tfrac{4}{9}\big((10^{2\cdot 1985}-1) - 2(10^{1985}-1)\big)$$
$$= \tfrac{4}{9}(10^{2\cdot 1985}-1) - \tfrac{8}{9}(10^{1985}-1)$$
$$= P - Q,$$

where P is the number consisting of a sequence of $2 \cdot 1985$ fours and Q consists of 1985 eights. Thus N consists of 1984 fours followed by 1 three followed by 1984 fives followed by 1 six. The sum of the digits in N is

$$1984 \cdot (4{+}5) + (6{+}3) = 1985 \cdot 9 = 17865.$$

Alternatively, one can replace $k = 1985$ by $k = 1, 2, 3$ and look for a pattern. When $k = 1$, $9ab = 360$ and the sum of the digits is 9. When $k = 2$, $9ab = 9 \cdot 88 \cdot 55 = 43560$ and the digit sum is 18. When $k = 3$, $9ab = 9 \cdot 888 \cdot 555 = 4435560$ and the digit sum is 27. It now seems clear (and can be proved by induction) that in general the digit sum is $9k$.

Query. We have shown that there is nothing special in this problem about 1985. Is there anything special about 8 and 5?

30. (E) Let $f(x) = 4x^2 - 40\lfloor x \rfloor + 51$ and let I_n be the interval $n \le x < n{+}1$ for integral n. Clearly $f(x) > 0$ for $x < 0$. For $x \ge 0$, $f(x)$ is increasing on each interval I_n since $4x^2$ is increasing and $-40\lfloor x \rfloor + 51 = -40n + 51$ is constant. Thus $f(x)$ has at most one root in each I_n and such a root will exist if and only if $f(n) \le 0$ and $f(n{+}1{-}\epsilon) > 0$ for small $\epsilon > 0$. Since $f(n{+}1{-}\epsilon)$ approaches $g(n) = 4(n{+}1)^2 - 40n + 51$ as ϵ approaches 0, it suffices to check if $g(n) > 0$ rather than checking $f(n{+}1{-}\epsilon)$ directly. Now,

$$f(n) \le 0 \iff (2n{-}3)(2n{-}17) \le 0$$
$$\iff \frac{3}{2} \le n \le \frac{17}{2}.$$

So it suffices to check I_n for $n = 2, 3, 4, 5, 6, 7, 8$. Checking we find $g(n) > 0$ for $n = 2, 6, 7, 8$; so there are four roots.

OR

Since $40\lfloor x \rfloor$ is even, $40\lfloor x \rfloor - 51$ is odd, implying that $4x^2$ must also be an odd integer, say $2k+1$, and $x = \sqrt{2k+1}/2$. Substituting in the original equation, it follows that

$$\left\lfloor \frac{\sqrt{2k+1}}{2} \right\rfloor = \frac{k+26}{20},$$

hence one must have $k \equiv 14 \bmod 20$. Furthermore,

$$\frac{k+26}{20} \leq \frac{\sqrt{2k+1}}{2} < \frac{k+26}{20} + 1.$$

Treating the two inequalities separately, multiplying by 20, squaring and completing the square, one obtains $(k-74)^2 \leq 70^2$ and $(k-54)^2 > 30^2$. Since x^2 must be positive, k is nonnegative, and it follows from the first inequality that $4 \leq k \leq 144$, and from the second one that either $k < 24$ or $k > 84$. Putting these together, one finds that either $4 \leq k < 24$ or $84 < k \leq 144$. In these intervals the only values of k for which $k \equiv 14 \bmod 20$ are $k = 14, 94, 114, 134$, yielding the four solutions

$$x = \frac{\sqrt{29}}{2}, \ \frac{\sqrt{189}}{2}, \ \frac{\sqrt{229}}{2}, \ \frac{\sqrt{269}}{2}.$$

Third Solution (from Prof. Paolo Pisaneschi, Pisa, Italy). We seek a graphical solution. The most direct approach would be to plot $y = 4x^2 - 40\lfloor x \rfloor + 51$ and see where it crosses the x-axis. However, $4x^2 - 40\lfloor x \rfloor + 51$ is an unfamiliar function, no easier to graph than to find zeros algebraically. The next standard approach is to break the function into two functions, each more familiar, and find where their graphs cross. Dividing the original equation by 40 and adding $\lfloor x \rfloor$ to both sides yields $\lfloor x \rfloor = \frac{1}{10}x^2 + \frac{51}{40}$, so we could find the solutions of

$$y = \lfloor x \rfloor$$
$$y = \frac{1}{10}x^2 + \frac{51}{40}.$$

However, this is still at best inconvenient, since both graphs are unbounded.

A better approach is to set $y = x - \lfloor x \rfloor$, which is bounded and periodic. Specifically, start with $0 = \frac{1}{10}x^2 - \lfloor x \rfloor + \frac{51}{40}$, add x to bold sides, substitute y for $x - \lfloor x \rfloor$ and rearrange, to obtain:

$$y = x - \lfloor x \rfloor$$
$$y = -\frac{1}{10}x^2 + x - \frac{51}{40}.$$

(Conversely, every solution (x, y) of this system yields a different solution x to the original equation.) The figure below shows clearly that there are four solutions to this system.

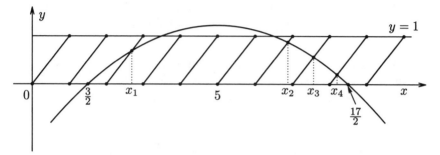

Furthermore, the graph indicates how to obtain the numerical values of the solutions: for $c = 2, 6, 7, 8$ solve $x - c = y = -\frac{1}{10}x^2 + x - \frac{51}{40}$, or $x^2 = (40c - 51)/4$. Thus the four solutions are the same as before.

1986 AHSME Solutions

1. (B) $[x - (y-z)] - [(x-y) - z] = (x - y + z) - (x - y - z)$
   $$= x - y + z - x + y + z = 2z.$$

2. (A) The slope of the original line is $2/3$ and the y-intercept is 4. So L has slope $1/3$ and y-intercept 8.

3. (D) Since $\angle C = 90°$ and $\angle A = 20°$, we have $\angle ABC = 70°$. Thus $\angle DBC = 35°$. It follows that $\angle BDC = 90° - 35° = 55°$.

4. (B) The statement S is false if there is a whole number n such that the sum of the digits of n is divisible by 6, but n itself is *not* divisible by 6. The number 33 has these properties.

5. (A)
   $$\left(\sqrt[6]{27} - \sqrt{6\tfrac{3}{4}} \right)^2 = \left[(3^3)^{1/6} - \left(\tfrac{27}{4}\right)^{1/2} \right]^2 = \left[\sqrt{3} - \frac{3\sqrt{3}}{2} \right]^2$$
   $$= \left[\frac{-\sqrt{3}}{2} \right]^2 = \frac{3}{4}.$$

6. (C) Let h, l and w represent the height of the table and the length and width of the wood blocks, respectively, in inches. From Figure 1 (copied from the exam) we have $l + h - w = 32$. From Figure 2 we have $w + h - l = 28$. Adding, $2h = 60$ and $h = 30$.

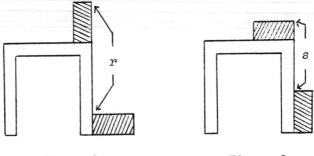

Figure 1   Figure 2

100

7 . (E) Let $\lfloor x \rfloor$ be the greatest integer less than or equal to x, and let $\lceil x \rceil$ be the least integer greater than or equal to x. If $x \leq 2$, then

$$\lfloor x \rfloor + \lceil x \rceil \leq 2 + 2 < 5,$$

so there are no solutions with $x \leq 2$. Similarly, if $x \geq 3$, then

$$\lfloor x \rfloor + \lceil x \rceil \geq 3 + 3,$$

so there are no solutions here. Finally, if $2 < x < 3$, then

$$\lfloor x \rfloor + \lceil x \rceil = 2 + 3 = 5,$$

so every such x is a solution. Thus the solution set is (E).

8. (E) With about 230 million people and under 4 million square miles, there are about 60 people per square mile. Since a square mile is about $(5000 \text{ ft})^2 = 25$ million square feet, that gives approximately $25/60$ of a million square feet per person. The closest answer is (E).

OR

Consider the approximation obtained by rounding to two significant digits, using scientific notation and simplifying. We obtain

$$\frac{(3.6 \times 10^6 \text{mi}^2)(5.3 \times 10^3 \text{ft/mi})^2}{2.3 \times 10^8 \text{people}} =$$

$$\frac{(3.6)(5.3)^2}{2.3} \times 10^{6+6-8} \text{ ft}^2/\text{person}.$$

Now, $3.6/2.3$ is slightly more than 1.5, and $(5.3)^2$ is between 25 and 30. Thus the exact answer is about $(28 \times 1.5)(10,000) = 420,000$. The closest answer to this among those given is (E).

9. (C) Factor each term of the given expression as the difference of two squares and group the terms according to signs to obtain

$$\left[\left(1-\frac{1}{2}\right)\left(1-\frac{1}{3}\right)\left(1-\frac{1}{4}\right)\cdots\left(1-\frac{1}{10}\right)\right]\left[\left(1+\frac{1}{2}\right)\left(1+\frac{1}{3}\right)\left(1+\frac{1}{4}\right)\cdots\left(1+\frac{1}{10}\right)\right]$$

$$= \left[\frac{1}{2}\frac{2}{3}\frac{3}{4}\cdot\frac{9}{10}\right]\left[\frac{3}{2}\frac{4}{3}\frac{5}{4}\cdots\frac{11}{10}\right] = \left[\frac{1}{10}\right]\left[\frac{11}{2}\right] = \frac{11}{20}.$$

OR

Note that

$$\left(1-\frac{1}{2^2}\right) = \frac{3}{4},$$

$$\left(1-\frac{1}{2^2}\right)\left(1-\frac{1}{3^2}\right) = \frac{3}{4}\cdot\frac{8}{9} = \frac{2}{3} = \frac{4}{6},$$

$$\left(1-\frac{1}{2^2}\right)\left(1-\frac{1}{3^2}\right)\left(1-\frac{1}{4^2}\right) = \frac{4}{6}\cdot\frac{15}{16} = \frac{5}{8}.$$

Clearly the pattern is

$$\left(1-\frac{1}{2^2}\right)\cdots\left(1-\frac{1}{n^2}\right) = \frac{n+1}{2n}.$$

(This may be proved by induction.) Thus the answer is $\frac{11}{20}$.

10. (E)   The first $24 = 4!$ words begin with A, the next 24 begin with E and the next 24 begin with H. So the 86th begins with M, and it is the $86 - 72 = 14^{\text{th}}$ such word. The first six words that begin with M begin with MA and the next six with ME. So the desired word begins with MH and it is the second such word. The first word that begins with MH is MHAES, the second is MHASE. Thus E is the letter we seek.

11. (B) $\triangle AHB$ is a right triangle. In any right triangle, the median from the right angle is half as long as the hypotenuse. (In short, much of the given information is irrelevant!)

12. (B)   Let c be the number John gets correct, w the number he gets wrong and u the number he leaves unanswered. That his score under the old scoring system would have been 84 means

$$30 + 4c - w = 84.$$

That his score under the new system is 93 means

$$5c + 2u = 93.$$

Since there are 30 problems,

$$c + w + u = 30.$$

Solving these three simultaneous equations gives

$$c = 15, \quad w = 6, \quad \text{and} \quad u = 9.$$

OR

We have

$$84 = 30 + 4c - w = (c+w+u) + 4c - w = 5c + u.$$

Subtracting $5c + u = 84$ from $5c + 2u = 93$, we obtain $u = 9$.

Note. The first sentence of this alternate solution shows that the old scoring system could have been described more simply as 5 times the number correct + the number left unanswered. The second sentence shows that the difference between the new and old systems is that the new system gives you an extra point for each unanswered question.

13. (E)   Since (4,2) is the vertex, $x = 4$ is the axis of symmetry. Since (2,0) is on the parabola, by symmetry so is (6,0). In other words, 2 and 6 are roots of $ax^2 + bx + c = 0$. Thus

$$y = ax^2 + bx + c = a(x-2)(x-6).$$

Substituting (4,2) we obtain

$$2 = a(2)(-2) = -4a.$$

So $a = -1/2$. Thus

$$y = -\frac{1}{2}x^2 + 4x - 6$$

and

$$abc = (-\frac{1}{2})(4)(-6) = 12.$$

OR

That (2,0) is on the graph means that $0 = 4a + 2b + c$. That (4,2) is on the graph means $2 = 16a + 4b + c$. That $x = 4$ is the axis of symmetry means $4 = -b/2a$. Thus we have 3 equations in the three unknowns a, b, c, which we may now solve.

14. (D)   We are given

$$b \text{ hops} = c \text{ skips}, \quad \text{or} \quad 1 \text{ hop} = \frac{c}{b} \text{ skips};$$

$$d \text{ jumps} = e \text{ hops}, \quad \text{or} \quad 1 \text{ jump} = \frac{e}{d} \text{ hops};$$

$$f \text{ jumps} = g \text{ meters}, \quad \text{or} \quad 1 \text{ meter} = \frac{f}{g} \text{ jumps}.$$

Substituting repeatedly gives

$$1 \text{ meter} = \frac{f}{g} \text{ jumps} = \frac{f}{g}\Big(\frac{e}{d} \text{ hops}\Big) = \frac{f}{g}\Big(\frac{e}{d}\Big(\frac{c}{b} \text{ skips}\Big)\Big) = \frac{cef}{bdg} \text{ skips}.$$

15. (C)   The true average A is $(x+y+z)/3$. The student computed

$$B = \frac{\dfrac{x+y}{2} + z}{2} = \frac{x+y+2z}{4}.$$

Thus

$$B - A = \frac{2z - x - y}{12} = \frac{(z-x) + (z-y)}{12},$$

which is always positive since $z > x$ and $z > y$.

Note. In effect, the student averaged the four numbers x, y, z, z. Since the largest number was counted twice, the student's average is larger than the average of x, y, z.

16. (C)   By the similarity of the two triangles,

$$\frac{PA}{PB} = \frac{PC}{PA} = \frac{CA}{AB};$$

hence

$$\frac{PA}{PC+7} = \frac{PC}{PA} = \frac{6}{8},$$

yielding the two equations

$$6(PC + 7) = 8PA \quad \text{and} \quad 6PA = 8PC.$$

From these one obtains $PC = 9$.

17. (B) For any selection, at most one sock of each color will be left unpaired, and this happens if and only if an odd number of socks of that color is selected. Thus 24 socks suffice: at most 4 will be unpaired, leaving at least 20 in pairs. However, 23 will do! Since 23 is not the sum of four odd numbers, at most 3 socks out of 23 will be unpaired. On the other hand, 22 will *not* do: if the numbers of red, green, blue and black socks are 5,5,5,7, then four are unpaired, leaving 9 pairs. Thus 23 is the minimum.

OR

Proceed inductively. If we require only one pair, then it suffices to select 5 socks. Moreover, selecting 4 socks doesn't guarantee a pair since we might select one sock of each color.

If we require two pairs, then it suffices to select 7 socks: any set of 7 socks must contain a pair; if we remove this pair, then the remaining 5 socks will contain a second pair as shown above. On the other hand, 6 socks might contain 3 greens, 1 black, 1 red and 1 blue — hence only one pair. Thus 7 socks is the smallest number to guarantee two pairs.

Similar reasoning shows that we must draw 9 socks to guarantee 3 pairs, and in general, $2p + 3$ socks to guarantee p pairs. This formula is easily proved by mathematical induction. Thus 23 socks are needed to guarantee 10 pairs.

18. (E) We claim that the minor axis has length 2, in which case the major axis has length $2 + .5(2) = 3$. To prove the claim, we may assume that the equation of the cylinder is $x^2 + y^2 = 1$ and that the intersection of the plane and the axis of the cylinder is $(0,0,0)$. By symmetry, $(0,0,0)$ is also the center of the ellipse. Furthermore, the line of intersection of the given plane with the x-y plane contains two points on the ellipse (and the cylinder), and they have $z = 0$. Finally, each "radius" of the ellipse extends from $(0,0,0)$ to some point (x, y, z) on the cylinder and thus has length

$$\sqrt{x^2 + y^2 + z^2} = \sqrt{1 + z^2}.$$

This has minimum value 1 when $z = 0$, which is obtained for the two diametrically opposed points described earlier. Thus the minor axis does indeed have length 2.

Note. For those with good spatial intuition, it should be evident that the minor axis of the ellipse is a diameter of the cylinder.

19. (A) Consider the figure. We may assume that Alice starts at A and ends at B. Observe that in $\triangle ABC$, $\angle ACB$ is right and AC has length $2\sqrt{3}$. The Pythagorean Theorem shows that $(AB)^2 = 13$.

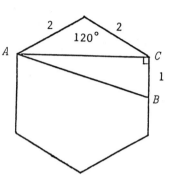

OR

Consider the second figure. Observe that in $\triangle ABD$

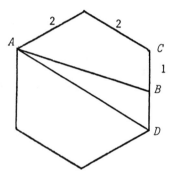

    1) $\angle ADB = 60°$
    2) $AD = 4$,
    3) $BD = 1$.

By the Law of Cosines,

$$(AB)^2 = 1^2 + 4^2 - 2(1)(4)(1/2) = 13.$$

Therefore $AB = \sqrt{13}$.

20. (E) To say that x and y are inversely proportional is to say that if x is multiplied by k, then y is divided by k. Let x' and y' be the new values after x increases by $p\%$. Then

$$x' = \left(1 + \frac{p}{100}\right)x \quad \text{and} \quad y' = \frac{y}{\left(1 + \frac{p}{100}\right)} = \frac{100}{100 + p}\,y.$$

By definition, the percentage decrease in y is

$$100\left(\frac{y - y'}{y}\right) = 100\left(1 - \frac{100}{100 + p}\right) = \frac{100p}{100 + p}.$$

Note. It is a common error to think that the correct answer to this question is (A). For instance, this error appeared in the early 1980s in the president's annual report from a famous American university!

21. (B)   The area of the shaded sector is $\frac{\theta}{2}(AC)^2$. This must equal half the area of $\triangle ABC$, which is $\frac{1}{2}(AC)(AB)$. Hence the shaded regions have equal area iff

$$\frac{\theta}{2}(AC)^2 = \frac{1}{4}(AC)(AB),$$

which is equivalent to

$$2\theta = \frac{AB}{AC} = \tan\theta.$$

22. (C)   There are $\binom{10}{6} = 210$ different sets which could be picked. If the second smallest pick is 3, then one number is picked from $\{1, 2\}$ and four are picked from $\{4, 5, \ldots, 9, 10\}$. There are $\binom{2}{1}\binom{7}{4} = 70$ ways to do this. So the probability is $70 \div 210 = 1/3$.

23. (E)   By the Binomial Theorem, $N = (69+1)^5 = (2 \cdot 5 \cdot 7)^5$. Thus a positive integer d is a factor of N iff $d = 2^p 5^q 7^r$, where p, q, r are each one of the 6 integers 0,1,2,3,4,5. Therefore there are $6^3 = 216$ choices for d.

24. (D)   Because $p(x)$ is a factor of

$$x^4 + 6x^2 + 25 \quad \text{and of} \quad 3x^4 + 4x^2 + 28x + 5,$$

it is also a factor of

$$3(x^4+6x^2+25) - (3x^4+4x^2+28x+5),$$

which equals

$$14x^2 - 28x + 70 = 14(x^2-2x+5).$$

Therefore, $p(x) = x^2 - 2x + 5$, and $p(1) = 4$.

OR

The unique complete factorization of $x^4 + 6x^2 + 25$ is a difference of squares:

$$x^4 + 10x^2 + 25 - 4x^2 = (x^2+5+2x)(x^2+5-2x).$$

Thus $p(x)$ is either the first quadratic or the second. Long division shows that only the second quadratic is a factor of $3x^4 + 4x^2 + 28x + 5$. Thus $p(x)$ is the second quadratic.

Note. It was not necessary to assume that b and c are integers. Even if they are allowed to be arbitrary complex numbers, $p(x)$ must still be $x^2 - 2x + 5$. The first solution shows this. (Why?) The second does not. (Why?) Can the second solution be modified so that it does show this?

25. (B)

$$\lfloor \log_2 N \rfloor = \begin{cases} 1 & \text{for } 2 \le N \le 2^2 \\ 2 & \text{for } 2^2 \le N \le 2^3 \\ \vdots & \qquad \vdots \\ 9 & \text{for } 2^9 \le N \le 2^{10} \\ 10 & \text{for } N = 2^{10} \end{cases}$$

Thus the desired sum is

$$1(2^2-2) + 2(2^3-2^2) + 3(2^4-2^3) + \cdots + 9(2^{10}-2^9) + 10$$

$$= 9 \cdot 2^{10} - (2^9+2^8+2^7+\cdots+2) + 10$$

$$= 9 \cdot 2^{10} - (2^9+2^8+2^7+\cdots+2+1) + 11$$

$$= 9 \cdot 2^{10} - (2^{10}-1) + 11$$

$$= 8 \cdot 2^{10} + 12 = 8(1024) + 12 = 8204,$$

where we have used the sum formula for a geometric series to obtain the next to last line.

Note. See the related AIME problem #8 from 1986.

26. (C) In *any* right triangle with legs parallel to the axes, one median to the midpoint of a leg has slope 4 times that of the other. This is easily shown by analytic geometry: any triangle of this sort may be labelled as in the figure, where a, b, c, d are arbitrary except $c \neq 0$, $d \neq 0$. (The figure has the right angle in the lower right, but the labelling allows complete generality — why?) Note that the slopes of the medians are

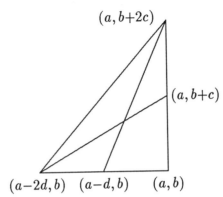

$$\frac{c}{2d} \quad \text{and} \quad \frac{2c}{d} = 4\left(\frac{c}{2d}\right).$$

Therefore, in our problem m is either 12 or 3/4.

In fact, both values are possible, each for infinitely many triangles. We show this for $m = 12$. Take any right triangle having legs parallel to the axes and a hypotenuse with slope 12/2, e.g., the triangle with vertices (0,0), (1,0), (1,6). Then the medians to the legs have slopes 12 and 3 (Why?). Now translate the triangle (don't rotate!) so that its medians intersect at the point where $y = 12x + 2$ and $y = 3x + 1$ intersect. This forces the medians to lie on these lines (Why?). Finally, for any central dilation of this triangle (a larger or smaller triangle with the same centroid and sides parallel to this one's sides), the medians will still lie on these lines.

Note. See the related AIME problem #15 from 1986.

27. (C) Because $AB \parallel DC$, $\text{arc } AD = \text{arc } CB$ and CDE and ABE are similar isosceles triangles. Thus

$$\frac{\text{Area } CDE}{\text{Area } ABE} = \left(\frac{DE}{AE}\right)^2.$$

Draw in AD. Since AB is a diameter, $\angle ADB = 90°$. Thus, considering right triangle ADE, $DE = AE \cos \alpha$, and

$$\left(\frac{DE}{AE}\right)^2 = \cos^2 \alpha.$$

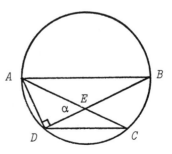

28. (C) Let s denote the length of a side of the pentagon. We compute the area of $ABCDE$ in two ways. First (Figure 1), it is the sum of the areas of the triangles OAB, OBC, OCD, ODE and OEA. Each of these has base s and altitude 1. Thus the area of the pentagon is $5s/2$. On the other hand (Figure 2), the area is the sum of the areas of the triangles ABC, ACD and ADE, which have base s and altitudes AQ, AP and AR, respectively. Thus the total area is

$$\frac{s}{2}\,(AP + AQ + AR).$$

Hence

$$AP + AQ + AR = 5 \quad \text{and} \quad AO + AQ + AR = 4.$$

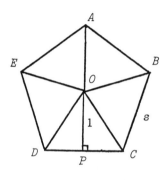
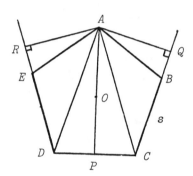

Figure .1   Figure 2

29. (B) Assume the altitude with length 4 falls on side a, the altitude with length 12 falls on b, and the unknown altitude with length h falls on c. Let K be the area of $\triangle ABC$. Then

$$4a = 12b = hc = 2K. \qquad (*)$$

By the triangle inequality, $c < a+b$ and $c > a-b$. In other words,

$$\frac{2K}{h} < \frac{2K}{4} + \frac{2K}{12} \quad \text{and} \quad \frac{2K}{h} > \frac{2K}{4} - \frac{2K}{12}\,,$$

These are equivalent to

$$\frac{1}{h} < \frac{1}{4} + \frac{1}{12} = \frac{1}{3} \quad \text{and} \quad \frac{1}{h} > \frac{1}{4} - \frac{1}{12} = \frac{1}{6}.$$

Hence $3 < h < 6$. If h is an integer, it cannot exceed 5.

OR

If the altitudes of a triangle are in a given ratio, the corresponding sides are in the inverse ratio. In particular, using the same notation as above, we have $a = 3b$ from $(*)$. Thus, thinking of side CB as fixed, A can be any point on the circle with center C and radius b as shown in the figure (except points X and Y,

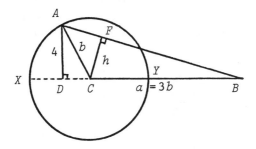

which don't result in a triangle). Since $\triangle FCB \approx \triangle DAB$,

$$\frac{h}{4} = \frac{3b}{AB}.$$

Furthermore,

$$AB < a+b = 4b$$

(AB would equal $4b$ when $A = X$) and

$$AB > a-b = 2b$$

(AB would equal $2b$ when $A = Y$). Thus

$$\frac{3b}{4b} < \frac{h}{4} < \frac{3b}{2b} \quad \text{or} \quad 3 < h < 6.$$

The largest integer value for h is 5.

This solution also suggests how to *construct* a triangle with altitudes 4, 12 and 5. From $(*)$, $c = 4a/5$. Thus A is an intersection point of the circle shown and the circle of radius $4a/5$ centered at B. There is only one hitch: this construction requires that you know a to begin with. Do you see how to get around this problem?

30. (B)   Either $x > 0$ or $x < 0$. Also, for any positive number a,

$$\frac{1}{2}\left(a + \frac{17}{a}\right) \geq \sqrt{17},$$

with equality only if $a = \sqrt{17}$, because this inequality is equivalent to $(a - \sqrt{17})^2 \geq 0$. Thus, if $x > 0$, then considering each of the given equations in turn, one deduces that $y \geq \sqrt{17}$, $z \geq \sqrt{17}$, $w \geq \sqrt{17}$ and $x \geq \sqrt{17}$. Suppose $x > \sqrt{17}$. Then

$$y - \sqrt{17} = \frac{x^2 + 17}{2x} - \sqrt{17}$$

$$= \left(\frac{x - \sqrt{17}}{2x}\right)(x - \sqrt{17}) < \frac{1}{2}(x - \sqrt{17}),$$

so that $x > y$. Similarly, $y > z$, $z > w$, and $w > x$, implying $x > x$, an obvious contradiction. Therefore $x = y = z = w = \sqrt{17}$, clearly a solution, is the only solution with $x > 0$. As for $x < 0$, note that (x, y, z, w) is a solution if and only if $(-x, -y, -z, -w)$ is a solution. Thus $x = y = z = w = -\sqrt{17}$ is the only other solution.

Alternative Solution (from Chan-Nhu Trinh, then a student, Collinsville IL). As in the first few lines above, we determine that if $x > 0$, then x, y, z, w are all $\geq \sqrt{17}$. Thus

$$\frac{1}{x} \leq \frac{1}{\sqrt{17}}, \quad \text{so} \quad \frac{17}{x} \leq \sqrt{17} \leq x, \quad \text{or} \quad x - \frac{17}{x} \geq 0.$$

Similarly, $y - \frac{17}{y} \geq 0$, $z - \frac{17}{z} \geq 0$, $w - \frac{17}{w} \geq 0$. Now, adding all four equations in the problem, and subtracting $x + y + z + w$ from both sides, we obtain

$$\left(x - \frac{17}{x}\right) + \left(y - \frac{17}{y}\right) + \left(z - \frac{17}{z}\right) + \left(w - \frac{17}{w}\right) = 0.$$

Since each term in parentheses is ≥ 0, we conclude they are all exactly 0. But the only positive solution to $x = \frac{17}{x}$ (or $y = \frac{17}{y}$, etc.) is $\sqrt{17}$. Thus the only solution to the original problem with $x > 0$ is $x = y = z = w = \sqrt{17}$. By the reasoning in the last two sentences of the original answer, the only other solution is $x = y = z = w = -\sqrt{17}$.

1987 AHSME Solutions

1. (D) $(1+x^2)(1-x^3) = 1(1+x^2) - x^3(1+x^2)$
   $$= 1 + x^2 - x^3 - x^5.$$

2. (E) The perimeter of ABC is 9. We delete two segments of length 1, DB and EB, and add a segment of length 1, DE. ($DE = 1$ because $\triangle DBE$ is also equilateral, since $DB = BE$ and $\angle DBE = 60°$.) Thus the perimeter of $ADEC$ is $9-2+1 = 8$.

3. (C) 7, 17, 37, 47, 67 and 97 are primes; for instance, for 97 it is sufficient to check that it is not divisible by 2, 3, 5 and 7, which are the only primes less than its square root. On the other hand, 27, 57=3 × 19, 77 and 87=3 × 29 are not primes.

4. (B) By inspection, we see that the fraction is of the form

   $$\frac{8a + 8b + 8c}{a + b + c} = \frac{8(a+b+c)}{a + b + c} = 8.$$

   Or, compute

   $$\frac{2 + 1 + \frac{1}{2}}{\frac{1}{2^2} + \frac{1}{2^3} + \frac{1}{2^4}} = \frac{\frac{7}{2}}{\frac{1}{4} + \frac{1}{8} + \frac{1}{16}} = \frac{\frac{7}{2}}{\frac{7}{16}} = 8.$$

5. (B) Since 12.5% is 1/8, N must be at least 8. Since $50\% = 1/2 = 4/8$ and $25\% = 1/4 = 2/8$, $N = 8$ is the answer.

6. (A) From the figure,

   $x = 180 - [(y+u) + (z+v)]$
   $= 180 - (y+z) - (u+v)$
   $= 180 - (y+z) - (180-w)$
   $= w - y - z.$

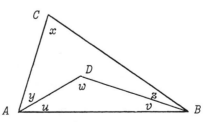

OR

Since $ADBC$ is a quadrilateral, the sum of its interior angles is $360°$. (It does not matter that the interior angle at D is a reflex angle.) Thus

$$x + y + z + (360 - w) = 360,$$
$$x = w - y - z.$$

7. (C) Adding 3 to each expression in the equality of the problem, we obtain

$$c = a + 2 = b + 5 = d + 7.$$

It is now easy to see that c is always the largest.

8. (C) Draw in DA and DB and drop perpendicular DE, forming rectangle $DCBE$. By the Pythagorean Theorem applied to triangles DCB and ADE, $DB = 5$ and $AD = \sqrt{116}$. Since $10 < \sqrt{116} < 11$, we find that $15 < AD + DB < 16$.

9. (B) The difference between the second and fourth terms is x; thus the difference between successive terms is $x/2$. Therefore $a = x/2$, $b = 3x/2$ and

$$\frac{a}{b} = \frac{x/2}{3x/2} = \frac{1}{3}.$$

10. (D) The simultaneous equations $a = bc$, $b = ca$, $c = ab$ imply

$$abc = (bc)(ca)(ab) = (abc)^2,$$

so either $abc = 0$ (ruled out) or $abc = 1$. The same simultaneous equations above also imply

$$abc = a^2 = b^2 = c^2,$$

so $|a| = |b| = |c| = 1$. It cannot be that all 3 unknowns are -1, nor can exactly one be -1, for in either case $a = bc$ is not satisfied. However, the remaining four cases,

$$(a, b, c) = (1, 1, 1), \ (-1, -1, 1), \ (-1, 1, -1) \text{ or } (1, -1, -1)$$

are all solutions. Thus there are 4 solutions in all.

11. (E)   Algebraic solution. Solving simultaneously, we obtain

$$x = \frac{5}{c+1}, \qquad y = \frac{3-2c}{c+1}.$$

We wish to find all values of c for which $x, y > 0$. First,

$$x > 0 \iff c+1 > 0 \iff c > -1.$$

Next, given that $c+1 > 0$, then

$$y > 0 \iff 3-2c > 0 \iff 3/2 > c.$$

Thus

$$x, y > 0 \iff -1 < c < 3/2.$$

OR

Geometric solution.   We wish to find those values of c for which the lines $x-y = 2$ and $cx+y = 3$ intersect inside Quadrant I. The line $x-y = 2$ has slope 1 and x-intercept 2, and is shown as L in the figure. The line $cx+y = 3$ has y-intercept 3 and slope $-c$. Several possible choices for this line are shown dashed in the figure. For this line to intersect L in Quadrant I, it is necessary and sufficient to choose a slope between that of L', which passes through $(2,0)$, and that of L'', which is parallel to L. Thus

$$\frac{-3}{2} < -c < 1,$$

or

$$-1 < c < \frac{3}{2}.$$

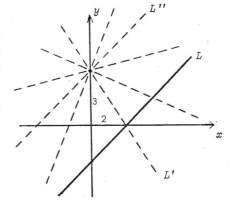

12. (D)   At any given time, the letters in the box are in increasing order from the bottom, because that's the way they were put in the box and the relative order of those still there is never changed. Thus, for any j and all $i < j$, all letters i taken off the pile later than letter j must still have been in the pile when j was added, and thus are in descending order from the top. Thus they must be taken off in descending order. For $j = 4$, this is not true in (D) – the lower numbered letters supposedly typed after 4 are 2,3,1, in that order. So (D) is not a possible typing order. All the other permutations listed meet the condition just described. For instance, for (B), the letters i with $i < 4$ which come after 4 are 3 and 1, in that order, which is descending order. All these other permutations are possible typing orders; the reader should show when each letter was added to the pile and when each was typed.

Note. This is a problem in computer science in disguise. The in-box is a "stack", an important "data structure" in computer science. The general question underlying this problem is: if items are put on a stack in order $1, 2, 3, \ldots, n$, then what permutations are possible orders for taking them off? We have found a necessary condition: all $i < j$ which follow j in the permutation must be in descending order. Is this condition sufficient?

Note 2. See the related AIME problem #15 from 1988.

13. (A)   Let $d_1, d_2, \ldots, d_{600}$ be the diameters of the concentric circles in the model. The d's form an arithmetic sequence with $d_1 = 2\,\text{cm}$ and $d_{600} = 10\,\text{cm}$. If L is the total length, then

$$L = \pi d_1 + \pi d_2 + \ldots + \pi d_{600}$$
$$= \pi(d_1 + d_2 + \cdots + d_{600}) = \pi 600 \left(\frac{d_1 + d_{600}}{2} \right)$$
$$= \pi 600 \left(\frac{12}{2} \right) \text{cm} = 36\pi \text{ meters.}$$

OR

Let L be the length of the tape in cm. The thickness of paper on the roll is $(10-2)/2 = 4$ cm. Therefore, the thickness of the tape is $4/600 = 1/150$ cm. We may assume that unfolding the paper and laying it out flat has negligible effect on its cross-sectional area. Therefore, we may equate the cross-sectional area of the laid out paper, which is $L/150$ cm^2, to the cross-sectional area while it is on the roll, which is

$$\pi 5^2 - \pi 1^2 = 24\pi \text{ cm}^2.$$

Solving for L gives $L = 3600\pi$ cm $= 36\pi$ meters.

14. (B) We may suppose that the sides of the square have length 2, so that $BM = ND = 1$. Then

$$\sin \theta = \sin(\frac{\pi}{2} - 2\alpha) = \cos 2\alpha$$
$$= 2\cos^2 \alpha - 1$$
$$= 2\left(\frac{2}{\sqrt{5}}\right)^2 - 1$$
$$= \frac{3}{5}.$$

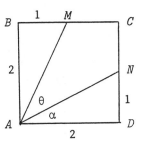

OR

One may express Area $\triangle AMN$ in terms of $\sin \theta$, find Area $\triangle AMN$ again numerically by subtracting the areas of other (right) triangles from the area of the square, and then solve for $\sin \theta$.

Third Solution. Since $\sin \theta = 3/5$, there must be a 3-4-5 triangle lurking somewhere. The following proof without words shows one place where it can be found. (This figure is an extension of the previous figure.)

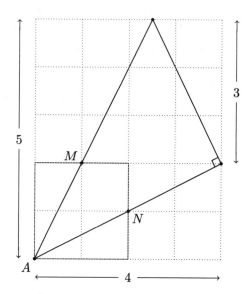

Remark. A whole article has appeared giving different solutions for this problem devised by the members of one school's math club. See Donald Barry, "An Abundance of Solutions", *The Mathematics Teacher*, Vol. 85, #5 (May 1992) 384–387.

15. (D)   One can solve for x and y individually, and then substitute, but it's a mess:

$$x = \frac{9 \pm \sqrt{57}}{2} \quad \text{and} \quad y = \frac{9 \mp \sqrt{57}}{2}.$$

It's sufficient, and easier, to solve for $x + y$:

$$
\begin{aligned}
63 &= x^2 y + x y^2 + x + y \\
&= xy(x{+}y) + (x{+}y) = (6{+}1)(x{+}y),
\end{aligned}
$$

so $x + y = 9$ and

$$x^2 + y^2 = (x{+}y)^2 - 2xy = 81 - 12 = 69.$$

16. (D)   The fact that VVW follows VYX establishes $W \leftrightarrow 0$ and $X \leftrightarrow 4$. Since VYX follows VYZ, X immediately follows Z, and this establishes that $Z \leftrightarrow 3$. That VVW follows VYX also establishes that V follows Y, so $Y \leftrightarrow 1$ and $V \leftrightarrow 2$. Hence $XYZ \leftrightarrow 413_5 = 108_{10}$.

17. (E)   Let A, B, C and D represent the scores of Ann, Bill, Carol and Dick, respectively. Then

$$(1) \quad A + C = B + D,$$
$$(2) \quad A + B > C + D,$$
$$(3) \quad D > B + C.$$

Adding (1) and (2) shows $2A > 2D$ or $A > D$. Subtracting (1) from (2) gives $B-C > C-B$ or $B > C$. By (3), $D > B$ (since C is nonnegative), so $A > D > B > C$.

Note. All the conditions in the problem can be satisfied, for instance, with the scores $A = 7$, $B = 3$, $C = 1$ and $D = 5$.

18 . (D)   Let the length of the shelf be 1 unit, and let x and y denote the thicknesses of the algebra and geometry books, respectively. Then

$$Ax + Hy = 1, \quad Sx + My = 1 \quad \text{and} \quad Ex = 1.$$

Thus $E = 1/x$ and it is sufficient to solve the first two equations above for x. Multiply the first by M, the second by H, and subtract: $(AM-SH)x = M-H$. Thus

$$E = \frac{1}{x} = \frac{AM - SH}{M - H}.$$

Note. The problem asserted that A, H, S, M, E are positive integers. One possible set of integer values are $A = 6$, $H = 4$, $S = 3$, $M = 6$, $E = 12$, which yield $x = 1/12$ and $y = 1/8$.

Query. In the solution, have we used the given information that $y > x$? That the constants are distinct? If so, how?

19 . (B) Since

$$\sqrt{65} - \sqrt{63} = \frac{(\sqrt{65}-\sqrt{63})(\sqrt{65}+\sqrt{63})}{\sqrt{65} + \sqrt{63}} = \frac{2}{\sqrt{65} + \sqrt{63}},$$

and since $\sqrt{65}$ and $\sqrt{63}$ are each equal to about 8, the answer must be near 1/8. Specifically,

$$\sqrt{65} + \sqrt{63} > 7.5 + 7.5 = 15, \quad \text{so} \quad \sqrt{65} - \sqrt{63} < \frac{2}{15} \approx .1333.$$

Thus the answer is (A) or (B). To determine which, we must decide whether $\sqrt{65} + \sqrt{63}$ is larger or smaller than $8 + 8$, since $2/16 = .125$ exactly. In fact, it is smaller: this is the case $n = 64$ and $a = 1$ of the inequality

$$\sqrt{n+a} + \sqrt{n-a} < 2\sqrt{n},$$

valid whenever $0 < |a| \le n$, which one may verify by squaring. (Alternately, the fact that the graph of $y = \sqrt{x}$ is concave down shows geometrically that $\sqrt{64}$ is greater than the average of $\sqrt{63}$ and $\sqrt{65}$ — how?) Therefore, $\sqrt{65} - \sqrt{63} > 2/16$ and the answer is (B).

Note. $\sqrt{65} - \sqrt{63} \approx .125004$, as can be verified using the Binomial series or (after the test) a calculator.

20. (A) Using $\log a + \log b = \log ab$ repeatedly, we find that the sum is

$$P = \log_{10}\left[(\tan 1°)(\tan 2°) \cdots (\tan 45°) \cdots (\tan 89°)\right]$$

Moreover, $(\tan 1°)(\tan 89°) = 1$, $(\tan 2°)(\tan 88°) = 1$, and so on, because $\tan\theta \tan(90°-\theta) = \tan\theta \cot\theta = 1$ for all θ at which both $\tan\theta$ and $\cot\theta$ are defined. Thus

$$P = \log_{10}(\tan 45°) = \log_{10} 1 = 0.$$

21. (B) First observe that the legs of the triangle are twice as long as the sides of the first inscribed square, so the legs have length $2\sqrt{441} = 42$. Now, let x be the side of the second inscribed square, as in the figure to the right. Note that all triangles in the figure are similar isosceles right triangles. Thus

$$42 = x\left(\sqrt{2} + \frac{1}{\sqrt{2}}\right) = \frac{3x}{\sqrt{2}},$$

$$x = 14\sqrt{2},$$

$$x^2 = 196 \cdot 2 = 392.$$

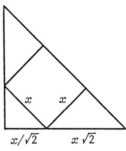

OR

The area of $\triangle ABC$ is $2 \cdot 441 = 882$. Subdivide the area of $\triangle ABC$ as shown on the right, and observe that 4 of the 9 congruent isosceles right triangles into which $\triangle ABC$ is partitioned are inside the square inscribed the second way. It follows that the area of that square is $(4/9)(882) = 392 \, \text{cm}^2$.

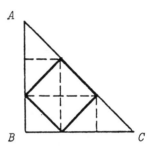

Note. See the related AIME problem #15 from 1987.

22. (C) In the figure, we show a cross section of the ball still in the ice. Since $\angle OCA = 90°$, $(r-8)^2 + 12^2 = r^2$. Solving, one finds

$$r = \frac{64 + 144}{16} = 13.$$

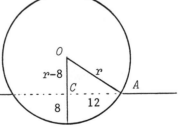

Query. We have assumed the ball floated with its center above the water, for otherwise the ball could not be removed without breaking the ice. Is there a second mathematical solution with the center below the water line?

23 . (D)   Applying the quadratic formula to $x^2 + px - 444p = 0$,   we find that its discriminant, $p(p+1776)$, must be a perfect square. This implies that $p + 1776$, and hence 1776, must be a multiple of p. Since $1776 = 2^4 \cdot 3 \cdot 37$,   it follows that $p = 2$, 3 or 37. It is easy to see that neither $2(2+1776)$ nor $3(3+1776)$ is a perfect square; so $p = 37$, which indeed satisfies the conditions of the problem, leading to 111 and -148 as the two roots. Since $31 < 37 \leq 41$,   the correct answer is (D).

OR

Rewrite the equation as $x^2 = p(444-x)$ and observe that $p|x^2$ (i.e., p is a factor of the integer x^2). Since p is prime, it follows that $p|x$ and so $x = np$ for some integer n. Substituting, the equation becomes $n^2 p = 444 - np$,   from which $n(n+1)p = 444$. Since the factorization of 444 into primes is $3 \cdot 4 \cdot 37$, it is evident that $p = 37$, $n = 3$ or -4, and $x = 111$ or -148.

24. (B)   The only such polynomial is $f(x) = x^2$. To prove this, recall that by definition a polynomial of degree n is a function

$$f(x) \;=\; \sum_{k=0}^{n} a_k x^k \;=\; a_n x^n + a_{n-1} x^{n-1} + \cdots + a_1 x + a_0,$$

where $a_n \neq 0$. For a polynomial, the equation

$$f(x^2) = \big[f(x)\big]^2 = f\big(f(x)\big)$$

is thus

$$\sum_{k=0}^{n} a_k (x^2)^k = \left[\sum_{k=0}^{n} a_k x^k\right]^2 = \sum_{k=0}^{n} a_k \left(\sum_{j=0}^{n} a_j x^j\right)^k. \qquad (*)$$

Polynomials are identical $\iff$ they are identical term by term. The highest-power terms in the three expressions in $(*)$ are, respectively,

$$a_n x^{2n}, \quad a_n^2 x^{2n}, \quad a_n^{n+1} x^{n^2}.$$

Thus $2n = n^2$ and $a_n = a_n^2 = a_n^{n+1}$, so $n = 2$ (since $n = 0$ was ruled out) and $a_n = 1$ (since $a_n \neq 0$). The first equation in $(*)$ now becomes

$$x^4 + bx^2 + c = (x^2 + bx + c)^2. \qquad (**)$$

It follows that $b = 0$; otherwise, the right side of $(**)$ has a cubic term but the left doesn't. Then it follows that $c = 0$; otherwise $(x^2+c)^2$ has an x^2 term but $x^4 + c$ doesn't. So $f(x) = x^2$ is the only candidate. One easily checks that it satisfies $(*)$.

Query. If we included the polynomials of degree 0 (the constant functions), how many more solutions would there be?

25 . (C)   Since $A = (0,0)$ and $B = (36, 15)$, we know that base AB of $\triangle ABC$ has length $3\sqrt{12^2 + 5^2} = 39$. We must choose C so that the height of $\triangle ABC$ is minimum. The height is the distance to $C = (x_0, y_0)$ from the line AB. This line is $5x - 12y = 0$. In general, the distance from (x_0, y_0) to the line $ax + by = c$ is

$$\frac{|ax_0 + by_0 - c|}{\sqrt{a^2 + b^2}}.$$

So in this case the distance is $|5x_0 - 12y_0|/13$. Since x_0 and y_0 are integers, the smallest this expression could be is $1/13$. ($0/13$ is not possible, for then C would be *on* line AB and we would not have a triangle.) The value $1/13$ is achieved, for instance, with $C = (5,2)$ or $(7,3)$. Thus the minimum area is $(1/2)bh = (1/2)(39)(1/13) = 3/2$.

OR

The triangle with vertices (x_1, y_1), (x_2, y_2), (x_3, y_3) has area

$$\frac{1}{2}\text{abs}\begin{vmatrix} x_2 - x_1 & y_2 - y_1 \\ x_3 - x_1 & y_3 - y_1 \end{vmatrix},$$

Setting $(x_1, y_1) = A$, $(x_2, y_2) = B$, $(x_3, y_3) = C$, we have

$$\text{Area } ABC = \tfrac{1}{2}\text{abs}\begin{vmatrix} 36 & 15 \\ x_3 & y_3 \end{vmatrix}$$

$$= \tfrac{1}{2}|36y_3 - 15x_3| = \tfrac{3}{2}|12y_3 - 5x_3|.$$

If x_3, y_3 are integers, the least nonzero value $|12y_3 - 5x_3|$ could have is 1. Indeed, when $x_3 = 5$, $y_3 = 2$, it *is* 1. So $3/2$ is the minimum area.

OR

This problem can be solved using Pick's Theorem. See the Solution to problem 13 of the 1985 AHSME for a statement of the theorem and references.

26. (B)   Let x be the first number, $y = 2.5 - x$ the second. We solve the problem graphically. By running the number line for y backwards, we can put y directly over x as follows:

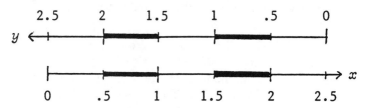

We have marked in bold the segments of their number lines where x and y are rounded up. We see that either they are both rounded up (in which case the resulting integers add to 3) or they are both rounded down (in which case the resulting integers don't add to 3). Since the bold segments for x have combined length 1 out of a total length of 2.5, and the probability distribution for x is uniform, the answer is $1/(2.5) = 2/5$.

Note. We haven't worried whether 1/2 and 3/2 are rounded up or down. Why doesn't it make any difference?

27. (B)   The cut $x = y$ separates the cheese into points with $x < y$ and those with $x > y$. Similarly for the other cuts. Thus, which piece a point is in depends only on the relative sizes of its coordinates, x, y, z. For instance, all points with $x < y < z$ are in the same piece. Since there are 3! ways to order x, y and z, there are 6 pieces.

OR

Since the planes $x = y$, $y = z$ and $z = x$ intersect along the line $x = y = z$, they divide all of space into just 6 pieces, not 8 the way most sets of 3 planes do. (Imagine forming the configuration by rotating a single half plane around that line; each rotation to the next position sweeps out one more piece.) Since the line $x = y = z$ is a major diagonal of the cheese, in each of the 6 pieces of space there is some point of the cheese. Thus the cheese is divided into 6 pieces also.

28. (D)   If $z = x+iy$ is a complex root of a polynomial equation with real coefficients, then so is $x - iy$. (In fact, if $x + iy$ is a multiple root, so is $x - iy$, with the same multiplicity. This is important below because in any sum involving roots it is meant that each root should be counted once for each multiplicity.) Now if $x + iy$ lies on the unit circle centered at $0 + 0i$, then the reciprocal of $x + iy$ is

$$\frac{x - iy}{x^2 + y^2} = x - iy.$$

Hence the reciprocal of each root is again a root. Thus the sum of the reciprocals of the roots is the same as the sum of the roots. In general, the sum of the roots of $a_n z^n + a_{n-1} z^{n-1} + \cdots + a_0$ is $-a_{n-1}/a_n$, so in our case the sum is simply $-a$.

Note. $-c/d$ is also a correct expression for the sum, but it wasn't listed. To see that it is correct, first let

$$f(z) = z^4 + az^3 + bz^2 + cz + d,$$
$$g(z) = dz^4 + cz^3 + bz^2 + az + 1.$$

Next observe that for $z \neq 0$

$$z^4 + az^3 + bz^2 + cz + d = z^4\left[1 + a\left(\frac{1}{z}\right) + b\left(\frac{1}{z}\right)^2 + c\left(\frac{1}{z}\right)^3 + d\left(\frac{1}{z}\right)^4\right].$$

Thus z is a nonzero root of $f(x) = 0$ iff $1/z$ is a root of $g(x) = 0$. Furthermore, all the roots of $f(x) = 0$ are nonzero, and hence their reciprocals exist, iff $d \neq 0$. (Why?) Thus, whenever all the roots of $f(x) = 0$ are nonzero (not just when they are on the unit circle), the sum of their reciprocals equals the sum of the roots of $g(x) = 0$. By the general fact stated in the last sentence of the solution, this latter sum is $-c/d$.

When the roots are on the unit circle, we can say more: $d = \pm 1$. (Why?) Thus one of c and $-c$ is also correct, but neither alone is always correct. [Exercises: With all roots on the unit circle, find a specific f with $d = 1$ and another with $d = -1$. For any such f show that the coefficients are either symmetric ($d = 1$, $c = a$) or antisymmetric ($d = -1$, $c = a$, $b = -b = 0$). Finally, extend everything in this Note to general n^{th} degree polynomials.]

29. (A)   It is easily seen by induction that $t_k > 1$ is true for all even k, and that $0 < t_k < 1$ is true for all odd $k > 1$. Hence n is odd, and t_{n-1} is $87/19$. Because $87/19 > 4$, it follows that $n - 1$ is divisible by 2 four times and that, subtracting 1 from t_{n-1} for each division by 2, term number $(n-1)/16$ is $(87/19) - 4 = 11/19$. Continuing in this way, we find that

$$\text{term number} \quad \frac{n-1}{16} - 1 = (n{-}17)/16 \quad \text{is} \quad 19/11,$$

and then

term number	$(n{-}17)/32$	is	$8/11$,
term number	$(n{-}49)/32$	is	$11/8$,
term number	$(n{-}49)/64$	is	$3/8$,
term number	$(n{-}113)/64$	is	$8/3$,
term number	$(n{-}113)/256$	is	$2/3$,
term number	$(n{-}369)/256$	is	$3/2$,
term number	$(n{-}369)/512$	is	$1/2$,
term number	$(n{-}881)/512$	is	1,
term number	$(n{-}881)/1024$	is	1.

Hence $(n{-}881)/1024$ equals 1, so $n = 1905$. The sum of the digits of n is 15.

Note. The given recursion is just a disguised form of the standard representation of a positive rational number in continued fraction form. See Chapter 1 of C. D. Olds, "Continued Fractions," Mathematical Association of America's New Mathematical Library, Vol. 9. Each positive rational number appears exactly once in the sequence given in the problem.

30. (E)   Point E is on AC, as in the original figure. To show this, we show that if $E = C$, then $\triangle ADE$ has more than half the area, hence DE is too far right. Indeed, in Figure 1 below, we may assume altitude CF is 1, in which case

$$\frac{\text{Area } EAD}{\text{Area } EAB} = \frac{AD}{AB} = \frac{1 + (1/\sqrt{3})}{1 + \sqrt{3}} = \frac{1}{\sqrt{3}} > \frac{1}{2}.$$

Thus we must move DE to the left, as in Figure 2, shrinking the dimensions of $\triangle EAD$ by a factor k so that

$$\text{Area } EAD = \tfrac{1}{2}\text{Area } CAB,$$

$$\tfrac{1}{2}k^2[1 + (1/\sqrt{3})] = \tfrac{1}{4}(1 + \sqrt{3}),$$

$$k^2 = \sqrt{3}/2,$$

$$k = \sqrt[4]{3/4}.$$

Thus

$$\frac{AD}{AB} = \frac{k[1 + (1/\sqrt{3})]}{1 + \sqrt{3}} = \frac{k}{\sqrt{3}} = \sqrt[4]{\tfrac{3}{4}}\sqrt[4]{\tfrac{1}{9}} = \frac{1}{\sqrt[4]{12}}.$$

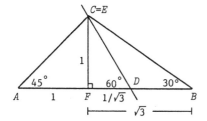

Figure 1

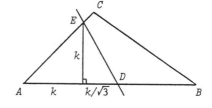

Figure 2

Alternative Solution. Again show that E is left of C by first assuming $E = C$ as in Figure 1 above. Apply the Law of Sines to $\triangle ABC$ and $\triangle ADC$, respectively:

$$\frac{AB}{\sin 105°} = \frac{AC}{\sin 30°}, \qquad \frac{AD}{\sin 75°} = \frac{AC}{\sin 60°}.$$

Since $\sin 105° = \sin 75°$, we obtain

$$\frac{AD}{AB} = \frac{\sin 30°}{\sin 60°} = \frac{1}{\sqrt{3}} > \frac{1}{2}.$$

Next, AD/AB is also the ratio of the areas of these triangles (when $C = E$), and if we shrink each linear dimension of $\triangle ADE$ by a factor k, we multiply the area by k^2. Thus we seek k so that $k^2/\sqrt{3} = 1/2$. Therefore $k = \sqrt[4]{3/4}$ and the answer to the problem is $k/\sqrt{3} = 1/\sqrt[4]{12}$.

1988 AHSME Solutions

1. (D)
$$\sqrt{8} + \sqrt{18} = \sqrt{4 \cdot 2} + \sqrt{9 \cdot 2}$$
$$= 2\sqrt{2} + 3\sqrt{2} = 5\sqrt{2}.$$

2. (D)   The ratios of corresponding sides are equal, so
$$\frac{XY}{AB} = \frac{YZ}{BC} \quad \text{or} \quad YZ = \frac{4 \cdot 5}{3} = 6\frac{2}{3}.$$

3. (A)   There are 4 strips, each of area 10. Also, 4 squares of side 1 are covered twice. So the total area covered is $4 \cdot 10 - 4 \cdot 1 = 36$.

4. (B)   Rewrite the equation as $y = -\frac{2}{3}x + 2$. With the equation in this form, the slope is the coefficient of x.

5. (E)   We have
$$x^2 + (2+b)x + 2b = x^2 + cx + 6,$$
so
$$c = 2 + b \quad \text{and} \quad 6 = 2b.$$
Thus $b = 3$ and $c = 5$.

6. (A)   A parallelogram has (among other things) 4 angles. "Equiangular" means that all angles are equal, so they are all 90°. Thus all equiangular parallelograms are rectangles. (They are not all squares since adjacent sides of a parallelogram need not be equal.) Conversely, every rectangle is a 4-sided, equiangular figure with opposite sides parallel. Thus a figure is an equiangular parallelogram if and only if it is a rectangle.

7. (D)
$$60 \text{ blocks} \times \frac{512 \text{ chunks/block}}{120 \text{ chunks/second}} =$$
$$256 \text{ seconds} \approx 4 \text{ minutes}.$$

Note. If we substitute "byte" for "chunk," we are talking standard terminology for communication between computers. A byte is a sequence of 8 "bits", each of which is a 0 or a 1. A transmission rate of 120 bytes/sec is called 1200 (not 120) "baud".

128

8. (B) $c = 3b = 3(2a) = 6a$, so

$$\frac{a+b}{b+c} = \frac{a+2a}{2a+6a} = \frac{3}{8}.$$

9. (C) Consider either diagonal of the table.   It has length $\sqrt{8^2+10^2} = \sqrt{164}$.   At some point in turning the table, this diagonal must be perpendicular to a pair of opposite walls of the room. Thus $S \geq \sqrt{164}$. This necessary condition is also sufficient: imagine a circle of diameter $\sqrt{164}$ within the room, translate the table so that it is inscribed in this circle, rotate the table $90°$ and then translate it into the desired corner. The smallest integer satisfying $S \geq \sqrt{164}$ is 13.

10. (D) From the given,

$$2.43553 \leq C \leq 2.44177.$$

Thus, rounded to one significant digit $C = 2$, to two digits $C = 2.4$, and to three digits $C = 2.44$ (not 2.43 even if $C = 2.43553$). However, rounded to four digits C might be as low as 2.436 and as high as 2.442. Thus the most precise value the experimenter can announce in significant digits is (D).

11. (C) By examining the data, we see that the percentage increases are

$$A : \frac{10}{40} = 25\% \quad B : \frac{20}{50} = 40\% \quad C : \frac{30}{70} = 42\tfrac{6}{7}\%$$

$$D : \frac{30}{100} = 30\% \quad E : \frac{40}{120} = 33\tfrac{1}{3}\%$$

Note. Greatest percentage increase is not the same as greatest absolute increase.

12. (A) From the units digits of the addition table of 1 through 9 (shown to the right) one sees that 0 appears 9 times and every other digit appears 8 times.

Note 1. This table is called the "mod 10" addition table. Usually a 0 row and a 0 column are also included.

	1	2	3	4	5	6	7	8	9
1	2	3	4	5	6	7	8	9	0
2	3	4	5	6	7	8	9	0	1
3	4	5	6	7	8	9	0	1	2
4	5	6	7	8	9	0	1	2	3
5	6	7	8	9	0	1	2	3	4
6	7	8	9	0	1	2	3	4	5
7	8	9	0	1	2	3	4	5	6
8	9	0	1	2	3	4	5	6	7
9	0	1	2	3	4	5	6	7	8

Note 2. Since there are 81 sums of two positive digits, the 10 units digits cannot be equally likely. This rules out (E).

13. (E) If $\sin x = 3\cos x$ then $\tan x = 3$. From the figure we conclude that

$$\sin x \cos x = \frac{3}{\sqrt{10}} \cdot \frac{1}{\sqrt{10}} = \frac{3}{10}$$

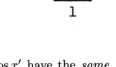

for any acute angle x. If x' is another angle with $\tan x' = 3$, then $x' - x$ is a multiple of π. Thus

$$\sin x' = \pm\frac{3}{\sqrt{10}}, \quad \cos x' = \pm\frac{1}{\sqrt{10}}.$$

So $\sin x' \cos x'$ is still $3/10$ (since $\sin x'$ and $\cos x'$ have the *same* sign).

OR

Multiplying the given equation first by $\sin x$ and then by $\cos x$ yields

$$\sin^2 x = 3\sin x \cos x,$$
$$cos^2 x = (1/3)\sin x \cos x.$$

Adding gives

$$1 = (10/3)\sin x \cos x,$$

so

$$\sin x \cos x = 3/10.$$

14. (A) The answer is

$$\frac{(-\tfrac{1}{2})(-\tfrac{3}{2})\cdots(-\tfrac{1}{2}-98)(-\tfrac{1}{2}-99)}{100!} \div \frac{(\tfrac{1}{2})(-\tfrac{1}{2})(-\tfrac{3}{2})\cdots(\tfrac{1}{2}-99)}{100!}$$

$$= \frac{(-\tfrac{1}{2}-99)}{\tfrac{1}{2}} = -199.$$

15. (A)   By long division, one finds that x^2-x-1 divides ax^3+bx^2+1 with quotient $ax + (a+b)$ and remainder $(2a+b)x + (a+b+1)$. But x^2-x-1 is a factor of ax^3+bx^2+1, so the remainder is 0. In other words,

$$2a + b = \ 0,$$
$$a + b = -1.$$

Solving, one obtains $a = 1,\ b = -2$.

OR

Since x^2-x-1 is a factor of ax^3+bx^2+1, the quotient must be $ax-1$ (why?). Thus

$$ax^3 + bx^2 + 1 = (ax-1)(x^2-x-1)$$
$$= ax^3 + (-a-1)x^2 + (1-a)x + 1.$$

Equating the coefficients of x^2 on the left and right, and then the coefficients of x, we obtain

$$b = -a - 1, \qquad 0 = 1 - a.$$

Hence $a = 1$ and $b = -2$.

Note. See the related AIME problem#13 from 1988.

16. (C)   Let $\triangle ABC$ and $\triangle A'B'C'$ have heights h and h'. The required ratio thus equals $(h'/h)^2$. Let O be the common center of the triangles and let M and M' be the intersections of BC and $B'C'$ with the common altitude from A. Since altitudes are also medians in equilateral triangles, $OM = h/3$ and $OM' = h'/3$. By hypothesis, $MM' = h/6$ so

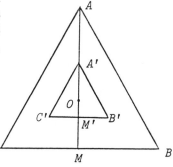

$$\frac{h}{3} = \frac{h'}{3} + \frac{h}{6}.$$

Thus

$$\frac{h'}{h} = \frac{1}{2} \quad \text{and} \quad \left(\frac{h'}{h}\right)^2 = \frac{1}{4}.$$

17. (C)   Label the equations

$$(1) \quad |x| + x + y = 10,$$
$$(2) \quad x + |y| - y = 12.$$

If $x \leq 0$, then (1) $\implies$ $y = 10$. Then (2) $\implies$ $x = 12$, a contradiction. Thus $x > 0$ and (1) becomes

$$(3) \quad 2x + y = 10.$$

If $y \geq 0$, (2) $\implies$ $x = 12$. Then (3) $\implies$ $y = -14$, a contradiction. Therefore $y < 0$ and equation (2) becomes

$$(4) \quad x - 2y = 12.$$

Solving (3) and (4) simultaneously, one finds $x = 32/5$, $y = -14/5$ and $x + y = 18/5$.

Alternatively, graph equations (1) and (2). The graph of (1) has two pieces: $y = 10$ for $x \leq 0$ and $2x + y = 10$ for $x \geq 0$. Each piece is easy to sketch. Similarly, (2) has two pieces: $x - 2y = 12$ for $y \leq 0$ and $x = 12$ for $y \geq 0$. A quick sketch of both equations shows that the only solution is in Quadrant IV, where the problem reduces to solving the linear equations (3) and (4).

18. (B)   There are 4 games in every play-off, and each game has 2 possible outcomes. For each sequence of 4 outcomes, the prizes are awarded a different way. Thus there are $2^4 = 16$ possible orders.

19. (B)   Let N be the numerator in the expression to be simplified. The answers given suggest that N is divisible by $bx + ay$. If the differing middle terms in the two parenthesized expressions in N, namely, $2a^2y^2$ and $2b^2x^2$, were not there, divisibility would be obvious. So let us separate the middle terms off. We obtain

$$N = bx(a^2x^2 + b^2y^2) + ay(a^2x^2 + b^2y^2) + bx(2a^2y^2) + ay(2b^2x^2)$$
$$= (bx + ay)(a^2x^2 + b^2y^2) + 2a^2bxy^2 + 2ab^2x^2y$$
$$= (bx + ay)(a^2x^2 + b^2y^2) + 2abxy(ay + bx).$$

Thus

$$\frac{N}{bx + ay} = a^2x^2 + b^2y^2 + 2abxy = (ax + by)^2.$$

20. (E)   The height of the rectangle is

$$XY = DE + FB = 2\sqrt{1^2 + 2^2} = 2\sqrt{5}\,.$$

The area of the rectangle is the same as the square's, i.e., 4. Therefore,

$$YZ = \frac{4}{2\sqrt{5}} = \frac{2}{\sqrt{5}} \quad \text{and} \quad \frac{XY}{YZ} = \frac{2\sqrt{5}}{2/\sqrt{5}} = 5.$$

21. (E)   Set $z = a + bi$. We seek $|z|^2 = a^2 + b^2$. So

$$z + |z| = 2 + 8i,$$
$$a + bi + \sqrt{a^2 + b^2} = 2 + 8i,$$
$$a + \sqrt{a^2 + b^2} = 2, \qquad b = 8,$$
$$a + \sqrt{a^2 + 64} = 2,$$
$$a^2 + 64 = (2-a)^2 = a^2 - 4a + 4,$$
$$4a = -60, \qquad a = -15.$$

Thus $a^2 + b^2 = 225 + 64 = 289.$

OR

In the complex plane, z and $|z|$ are vectors of equal length, and $z + |z| = 2 + 8i$ is their vector sum. So $0, z, 2+8i$, and $|z|$ form the vertices of a rhombus, as in the figure. The diagonals of a rhombus bisect each other and are perpendicular. The diagonal from 0 to $2+8i$ has slope 4 and midpoint $1+4i$. Thus the diagonal from z to $|z|$ passes through $1+4i$ with slope $-1/4$. Therefore this diagonal intersects the real axis at $x = 17$. Since $|z|$ is on the real axis, we conclude that $|z| = 17$ and $|z|^2 = 289$.

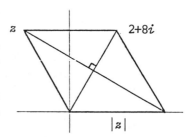

22. (A)   In any triangle with sides a, b, c, the angle opposite a is acute iff $a^2 < b^2 + c^2$. This follows from the Law of Cosines. Applying this fact in turn to the angle opposite x, 24 and 10, we find

$$x^2 < 10^2 + 24^2 = 26^2,$$
$$24^2 < x^2 + 10^2 \iff 476 < x^2,$$
$$10^2 < x^2 + 24^2.$$

The first line tells us that $x < 26$. The second tells us that $x \geq 22$ (since x is an integer). The third is satisfied for every x. Thus there are 4 integer values which meet all the conditions: 22, 23, 24, 25.

23. (B)   Consider the edge of length 7 and the two triangular faces of $ABCD$ which share this edge. For both of these triangles, the other two sides must have lengths differing by less than 7, for otherwise the Triangle Inequality would be violated. Of the numbers given, only the pairs (13,18) and (36,41) satisfy this requirement, leaving the edge of length 27 as the one opposite to that of length 7. Consequently, we must have one of the two arrangements pictured below:

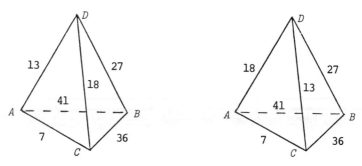

The arrangement on the left is impossible because $\triangle ABD$ fails to satisfy the Triangle Inequality. This leaves the arrangement on the right, in which $CD = 13$.

The reader may wish to verify that the second arrangement *is* possible by constructing a physical model — or by showing mathematically that it is constructible.

24. (C)   By viewing the sides of the trapezoid as tangents to the circle, we find that the sums of the lengths of opposite sides are equal. (Indeed, this is true for any circumscribed quadrilateral.) Defining x and y as shown in the figure, we have

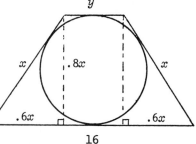

$$2y + 1.2x = 2x,$$

$$y + 1.2x = 16.$$

Then $y = 4$, $x = 10$, and the area is $\frac{1}{2}(4{+}16)(8) = 80$.

Alternative Solution. Consider the second figure, where lines have been drawn from the center of the circle to certain sides and a vertex. It is given that $\sin 2\alpha = .8$. Using the double angle formula

$$\tan \alpha = \frac{\sin 2\alpha}{1 + \cos 2\alpha},$$

we find

$$\tan \alpha = \frac{.8}{1 + .6} = .5.$$

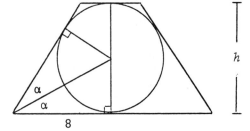

Thus $h/2 = 4$ or $h = 8$. We now know that the two 3–4–5 right triangles in the first figure each have base $\frac{3}{4} \cdot 8 = 6$. Thus $16 = 6{+}y{+}6$ or $y = 4$. Finally, the area is $\frac{1}{2}(4{+}16)(8) = 80$ as before.

25. (E)   Let the numbers of elements in the sets be given by $|X| = x$, $|Y| = y$ and $|Z| = z$. Then

$$|X{\cup}Y| = x + y, \qquad |X{\cup}Z| = x + z, \qquad \text{and} \qquad |Y \cup Z| = y + z.$$

The given information can be summarized in the following 3 equations in 3 unknowns:

$$\frac{37x + 23y}{x + y} = 29, \qquad \frac{37x + 41z}{x + z} = 39.5, \qquad \frac{23y + 41z}{y + z} = 33.$$

Simplifying these equations, we obtain $4x = 3y$, $5x = 3z$, $5y = 4z$. We want the value of the fraction

$$\frac{37x + 23y + 41z}{x + y + z}.$$

Making the substitutions $y = 4x/3$ and $z = 5x/3$, we obtain

$$\frac{37x + 23(\frac{4x}{3}) + 41(\frac{5x}{3})}{x + \frac{4x}{3} + \frac{5x}{3}} = \frac{111x + 92x + 205x}{3x + 4x + 5x} = 34.$$

Queries. Where did we use the given that X, Y and Z are disjoint? Also, did we need *all* the information which was given?

26. (D)   Let t be the common value of $\log_9(p)$, $\log_{12}(q)$ and $\log_{16}(p{+}q)$. Then

$$p = 9^t, \qquad q = 12^t, \quad \text{and} \quad 16^t = p + q = 9^t + 12^t.$$

Divide the last equation by 9^t and note that

$$\frac{16^t}{9^t} = \left(\frac{4^t}{3^t}\right)^2 = \left(\frac{12^t}{9^t}\right)^2 = \left(\frac{q}{p}\right)^2.$$

Now let x stand for the unknown ratio q/p. From the division referred to above we obtain $x^2 = 1 + x$, which leads easily to $x = \frac{1}{2}(1{+}\sqrt{5})$ since x must be the positive root.

27. (D) $ABCD$ is a trapezoid, hence its area is $BC(AB{+}CD)/2$. With an eye on the answers, we seek to express BC in terms of AB and CD. Let M be the point where BC is tangent to the circle, and let N be the point where AB intersects the circle. $\angle AND$ is right because AD is a diameter. Thus $BCDN$ is a rectangle and $BN = CD$. $\angle OMB$ is right because BC is a tangent. Thus OM is parallel to CD and M is the midpoint of BC. A standard theorem about tangents and secants to circles now states that $(BM)^2 = BN \cdot BA$, so that $BC = 2\sqrt{CD \cdot BA}$. Hence the area of $ABCD$ is $(AB{+}CD)\sqrt{AB \cdot CD}$. This is an integer iff $AB \cdot CD$ is a perfect square, since AB and CD are integers in all the choices given. Thus the answer is (D).

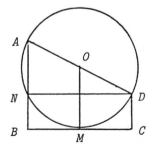

28. (D)   We must solve for p when

$$\binom{5}{3}p^3(1-p)^2 = 10p^3(1-p)^2 = \frac{144}{625},$$

or $p^3(1-p)^2 = 72/5^5$.   If we define

$$f(p) = p^3(1-p)^2 - \frac{72}{5^5},$$

we see that $f(0) = f(1) = -72/5^5$,   and

$$f(\tfrac{1}{2}) = \frac{1}{32} - \frac{72}{5^5} \approx .03 - .024 > 0.$$

Thus, since f is continuous, f has at least *two* real roots, r_1, r_2, satisfying $0 < r_1 < \tfrac{1}{2}$ and $\tfrac{1}{2} < r_2 < 1$. In fact, $r_1 = 2/5$ and $3/5 < r_2 < 4/5$.

29. (A)   The best fit line goes through some point (x_2, z). Claim: the correct slope for this line makes the directed vertical distances to it from (x_1, y_1) and (x_3, y_3) the same. To see this, imagine any line through (x_2, z) and rotate it until the two directed vertical distances to it are the same. Call this distance d. For any other line through (x_2, z) the directed distances to it from (x_1, y_1) and from (x_3, y_3) will be $d+e$ and $d-e$ for some $e \neq 0$. Thus the sum of the squares will increase from

$$2d^2 + (z-y_2)^2$$

to

$$(d+e)^2 + (d-e)^2 + (z-y_2)^2 = 2d^2 + 2e^2 + (z-y_2)^2.$$

This proves the claim.   Finally, the slope from (x_1, y_1+d) to (x_3, y_3+d) is $(y_3-y_1)/(x_3-x_1)$, whatever d and z are!

   Note. There are several formulas in statistics which give the slope of the best fit line for arbitrary data points. One can use any of these general formulas to solve this special case, but the algebra is quite involved.

   Queries.   What are d and z?   Where was the hypothesis $x_3-x_2 = x_2-x_1$ used?

30. (E)   Note that $x_0 = 0$ gives the constant sequence $0, 0, \ldots$, since $f(0) = 4 \cdot 0 - 0^2 = 0$.   Because $f(4) = 0$, $x_0 = 4$ gives the sequence $4, 0, 0, \ldots$ with two different values. Similarly, $f(2) = 4$ so $x_0 = 2$ gives the sequence $2, 4, 0, 0, \ldots$ with three values. In general, if $x_0 = a_n$ gives the sequence $a_n, a_{n-1}, \ldots, a_2, a_1, a_1, \ldots$ with n different values, and $f(a_{n+1}) = a_n$, then $x_0 = a_{n+1}$ gives a sequence with $n+1$ different values. (It could not happen that $a_{n+1} = a_i$ for some $i < n+1$; why?) Thus, it follows by induction that there is a sequence with n distinct values for every positive integer n — as soon as we verify that there *is* always a real number a_{n+1} such that $f(a_{n+1}) = a_n$. This follows from the quadratic formula: First, the solutions to

$$f(a_{n+1}) = 4a_{n+1} - a_{n+1}^2 = a_n$$

are

$$a_{n+1} = 2 \pm \sqrt{4 - a_n}.$$

Second, if $0 \le a_n \le 4$, then a_{n+1} is real; in fact, $0 \le a_{n+1} \le 4$ (why?). Third, $0 = a_1 \le 4$. Thus, by induction, all terms satisfy $0 \le a_n \le 4$; in particular, they are all real.

Note. This problem is actually an introduction to "Chaos", the study of the seemingly random behavior one can get from simple rules. The rule here is to iterate $x_n = f(x_{n-1})$. In this particular case, an exact formula can be found for an x_0 that has n distinct iterates. See David Sprows, Reiterating an AHSME Problem, *Mathematics and Computer Education*, Vol. 25 #1 (1991) 17–21.

Solutions for Dropped AHSME Problems

1. (E) As we show below, if $a \geq 0$ the only solution is $2a$; if $a \leq 0$ the solutions are $-10a/3$, $2a$ and $6a$. So none of the choices (A)–(D) is always correct.

First, for x to be a solution of

$$\sqrt{x^2 - 7ax + 10a^2} - \sqrt{x^2 + ax - 6a^2} = x - 2a,$$

means that the radicands (expressions under radical signs) are nonnegative and that the equation is true with the radicals representing the nonnegative square roots of the radicands. (When b is negative, $\sqrt{b}$ is not defined because there is no standard convention as to whether $\sqrt{b}$ means $+i\sqrt{-b}$ or $-i\sqrt{-b}$.)

Factoring the radicands, the equation becomes

$$\sqrt{(x-5a)(x-2a)} - \sqrt{(x-2a)(x+3a)} = x - 2a. \qquad (1)$$

Thus $x = 2a$ is a root (both sides equal 0). If $x \neq 2a$, either $x-2a > 0$ or $x-2a < 0$. In the former case, we may divide (1) by $\sqrt{x-2a}$ and use the rule $\sqrt{cd} = \sqrt{c}\sqrt{d}$ (correct for $c, d \geq 0$) to obtain

$$\sqrt{x-5a} - \sqrt{x+3a} = \sqrt{x - 2a}. \qquad (2)$$

If $x - 2a < 0$, we may divide (1) by $\sqrt{-(x-2a)}$ to obtain

$$\sqrt{-(x-5a)} - \sqrt{-(x+3a)} = -\sqrt{-(x - 2a)}. \qquad (3)$$

Squaring (2), which may introduce extraneous roots, yields

$$(x - 5a) - 2\sqrt{(x-5a)(x+3a)} + (x + 3a) = x - 2a$$
$$2\sqrt{(x-5a)(x+3a)} = x. \qquad (4)$$

Squaring (3) yields

$$-(x - 5a) - 2\sqrt{(x-5a)(x+3a)} - (x + 3a) = -(x - 2a)$$
$$-2\sqrt{(x-5a)(x+3a)} = x. \qquad (5)$$

From either (4) or (5), squaring yields the same thing, so we are back to one case:

139

$$4(x^2 - 2ax - 15a^2) = x^2,$$
$$3x^2 - 8ax - 60a^2 = 0,$$
$$(3x+10)(x-6a) = 0.$$

Thus $x = -\frac{10}{3}a$ and $x = 6a$ are the only other solution candidates. Plugging the first of these into (1) we obtain

$$\sqrt{\left(-\frac{25}{3}a\right)\left(-\frac{16}{3}a\right)} - \sqrt{\left(-\frac{16}{3}a\right)\left(-\frac{1}{3}a\right)} = -\frac{16}{3}a,$$

$$\frac{20}{3}|a| - \frac{4}{3}|a| = -\frac{16}{3}a,$$

$$\frac{16}{3}|a| = -\frac{16}{3}a. \tag{6}$$

Clearly, (6) is true iff $a \le 0$. Similarly, plugging $x = 6a$ into (1) reduces to

$$2|a| - 6|a| = 4a,$$

and this also is true iff $a \le 0$.

Query. A common approach to this problem would be first to divide (1) by $\sqrt{x - 2a}$ to obtain

$$\sqrt{x-5a} - \sqrt{x+3a} = \sqrt{x - 2a}. \tag{7}$$

and then to solve this by squaring twice. If you are not careful, you get that the roots to (7) are $x = -\frac{10}{3}a, 6a$. What is the correct solution to (7)? Why is this different from the correct solution to (1)?

2. (B)   Near a root r of any polynomial $g(x)$, the graph looks like some constant times the highest power of $x - r$ that divides $g(x)$; the other factors (whose product changes very little in value for x near r) determine the constant, and all that is really important is the constant's sign. For instance, near $x = -2$, the $g(x)$ of this problem looks like $c(x+2)^2$, where c is the product of the other factors, $(x-2)^2(x-1)^3(x+3)^2$ at $x = -2$. This product is negative, since the cubic factor is negative; thus $g(x)$ looks like a parabola opening down near $x = -2$. Considering all four roots, one finds that $g(x)$ looks like

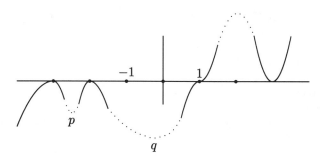

Point p is the lowest point between the roots at -3 and -2; q is the lowest point between the roots at -2 and 1. How low these points are cannot be determined without much more work, but so long as ϵ is much smaller than m, the lesser of these two heights, the graph of $h(x)$ looks like

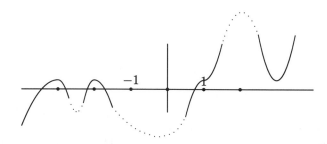

Thus $h(x)$ has 5 real roots. (In fact, ϵ merely has to be smaller than m, not much smaller.)

3. (A)   Since $c \neq 0$, zero is not a root of the given equation, and the given equation is equivalent to the equation $a + b(1/x) + c(1/x)^2 = 0$. Therefore

$$\frac{1}{x} = \frac{-b \pm \sqrt{b^2 - 4ac}}{2c}$$

$$x = \frac{2c}{-b \pm \sqrt{b^2 - 4ac}}.$$

OR

$$x = \frac{-b \pm \sqrt{b^2 - 4ac}}{2a} \cdot \frac{-b \mp \sqrt{b^2 - 4ac}}{-b \mp \sqrt{b^2 - 4ac}} = \frac{2c}{-b \mp \sqrt{b^2 - 4ac}}.$$

Remark. This problem was deleted in part because it was felt to be too tricky; students would jump at what looked like the traditional formula. But in fact, the alternative formula in this problem is of computational importance. Suppose b is large and positive but $4ac$ is small, positive or negative. Then the numerator in $(-b + \sqrt{b^2 - 4ac})/2a$ may be very difficult to compute accurately due to roundoff error. But the equivalent form $2c/(-b - \sqrt{b^2 - 4ac})$ suffers from no such problem, since you take the sum of two large quantities instead of their difference. (For further discussion of this alternative formula, see Anthony Ralston, "The Really New College Mathematics and Its Impact on the High School Curriculum," in *The Secondary School Mathematics Curriculum,* 1985 Yearbook of the National Council of Teachers of Mathematics, Christian Hirsch, ed., NCTM, Reston VA, 1985.)

4. (B)   The given equations may be rewritten as

$$c = \frac{a + e}{2}, \qquad c = \frac{b + d}{2}, \qquad d = \frac{c + e}{2},$$

so c is halfway between a and e on the number line, c is also halfway between b and d, and d is halfway between c and e. Without loss of generality we may assume $a < c < e$. (The other case is obtained by taking the mirror image around c, and this will not change the set consisting of both the largest and smallest from a, b, c, d, e.) Next, either $b < c < d$ or $d < c < b$, but in the second case d could not be between c and e. Therefore, $a < b < c < d < e$. ($a < b$ because $d < e$ and c is exactly halfway between a, e and between b, d.) Thus $\{a, e\}$ is the set of the largest and smallest.

5. (A) If the equation is $ax^2 + bx + c = 0$, the discriminant is $b^2 - 4ac$. If b is even, the discriminant is divisible by 4. If $b = 2m+1$ (b is odd), the discriminant is $4m^2 + 4m + 1 - 4ac$, so it has remainder 1 when divided by 4. In other words, for an integer D to be the discriminant of a quadratic equation with integer coefficients, it is necessary that $D \equiv 0, 1 \pmod 4$. Thus 23 cannot be a discriminant.

To show that the other answers *can* be discriminants, one may find appropriate a, b, c by trial and error. For instance, $24 = 6^2 - 4(1)(3)$. Better yet, show that the condition $D \equiv 0, 1 \pmod 4$ is also *sufficient* to make D a discriminant. Hint: it suffices to consider $a = 1$ and $b = 0, 1$.

6. (D) Basically, one must try cases, so the "trick" is to think of an organizational principle that eliminates most ten-digit numbers right away. One such principle (found after much experimentation) is by the number of zero digits.

Let N be any number with this property, and let $d_1, d_2, \ldots, d_{10}$ be its digits from left to right. By the definition of N, the sum of its digits equals the number of digits, 10. Let z be the number of zeros in N. Then $d_1 = z$, and there are $10 - z - 1$ other nonzero digits. Since their sum is $10 - z$, these $9 - z$ other digits must consist of $8-z$ 1's and one 2. Further, the number of nonzero digits in N must equal the number of distinct digits in N (since $d_k \geq 1$ iff $k-1$ appears at least once). So if $z > 2$, N has 4 distinct digits, $0, 1, 2$ and z; otherwise N has 3 distinct digits, $0, 1, 2$. But N also has $10 - z$ nonzero digits. Thus, $z = 6$ and the digits must be $1, 1, 2, 6, 0, 0, 0, 0, 0, 0$. By the definition of N, they must appear in the order 6,210,001,000. Thus there is exactly one number N.

Query. Suppose we don't require that the number be written in base 10. Does the answer change?

7. (A) Let v be the speed of
the bus in mph. Suppose
the runner could catch the
bus by going to P in the
figure. Thus it takes the bus
equal time or longer to reach
P from B than it takes the
runner to reach P from G.
That is, $BP/v \geq GP/15$, or

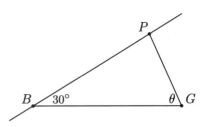

$v \leq 15\,BP/GP$. Since v is the largest number for which this is
possible (that is, there is some point P the runner can run to
and catch the bus), then $v = 15\,BP_0/GP_0$, where P_0 is the point
on the line BP for which BP/GP is maximum. By the Law of
Sines,

$$\frac{BP}{GP} = \frac{\sin\theta}{\sin 30°} = 2\sin\theta,$$

where $\theta = \angle BGP$. Thus $BP_0/GP_0 = 2\sin 90° = 2$ and $v = 15\cdot 2 =$
30.

Note. The runner should *not* go to the closest point on the
road (for which BP/GP is only $\sqrt{3}$). Rather, she should always
run perpendicularly to the direction in which she first saw the
bus – in this case she should run due north. Thus, even if she
does not know the speed of the bus or the angle of the road, her
optimal strategy is the same: if she can catch the bus at all, she
can catch it by running perpendicularly! (What does depend on
the direction of the road is the maximum speed of a bus she can
catch.) Though not many people have to run over flat country to
catch buses, historically many military ships have had to intercept
other ships, and the rule "head perpendicular to the sighting" is
well known to naval officers.

8. (C)   Let x_0 be any solution to $f_{1981}(x) = f_{1982}(x)$ and let $x_n = f_n(x_0)$. Then the given equation may be rewritten as $x_{1981} = |1 - x_{1981}|$, so $x_{1981} = 1/2$. Since $x_{1981} = |1 - x_{1980}|$, it follows that $x_{1980} = 1/2, 3/2$. In general, $c = |1 - x|$ has two solutions for x, namely $1 - c$ and $1 + c$, so long as $c \geq 0$; see the figure. Next, x_{1979} has 4 possible values, $1/2$ and $3/2$ from $x_{1980} = 1/2$, and $-1/2$ and $5/2$ from $x_{1980} = 3/2$. However, x_{1979} is an output of f, so it must be nonnegative. Continuing in this way, $x_{1978} = 1/2$, $3/2$, $5/2$, $7/2$, and for each $k > 1$, x_{k-1} has one more solution than x_k. Thus $x_1 = j + \frac{1}{2}$, where $j = 0, 1, ..., 1981$. Finally, x_0 has no sign restrictions, so the number of solutions for x_0 is $2 \cdot 1981 = 3962$.

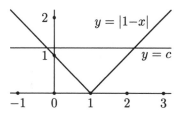

OR

If the output of $f(x)$ is nonnegative, then

$$|1 - f(x)| = \begin{cases} 1 - f(x) & \text{if } 0 \leq f(x) \leq 1, \\ f(x) - 1 & \text{if } f(x) \geq 1. \end{cases}$$

Thus for $n \geq 1$ the graph of $f_{n+1}(x)$ may be obtained from the graph of $f_n(x)$ by 1) taking the part with height at most 1 and reflecting it around the horizontal line at $y = 1/2$; and 2) shifting the rest down one unit. Starting with $f_1(x)$ (shown previously) and iterating, we find that the graphs of $f_n(x)$ and $f_{n+1}(x)$ are as shown below. Therefore, f_n and f_{n+1} cross at every half-integer from $x = \frac{3}{2} - n$ through $x = n + \frac{1}{2}$. Thus there are $2n$ solutions to $f_{n+1}(x) = f_n(x)$. For $n = 1981$ there are 3962 solutions.

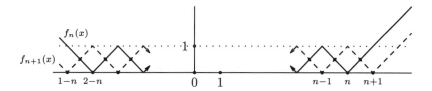

9. (B)   Let s be the side length of $\triangle ABC$. $\triangle AED$ is 30–60–90 with AE the shortest side. Since $AE = 1$, then $DE = \sqrt{3}$ and $AD = 2$. Now, $DECB$ circumscribes the circle, so $DE + BC = EC + BD$. (This theorem follows from the fact that two tangents to the same circle from the same external point have the same length.) Thus $\sqrt{3} + s = (s{-}1) + (s{-}2)$, and so $s = 3 + \sqrt{3}$.

OR

Let the circle have radius r. Consider the figure below, where T and Q are the points of tangency on DE and CE respectively, and radii OT, OQ have been drawn. $\angle OTE = \angle TEQ = \angle EQO = 90°$. Thus $OTEQ$ is a rectangle. (In fact, it is a square, since $OT = OQ = r$, but we don't need this.) Thus $AQ = 1{+}r$. But Q is the midpoint of AC, since $\triangle ABC$ is equilateral. So $1{+}r = s/2$. Because $\triangle ABC$ is equilateral, $r = s\sqrt{3}/6$ (O is at the centroid and the altitude is $s\sqrt{3}/2$). Thus $s\sqrt{3}/6 = (s/2) - 1$, from which algebra leads to $s = 3 + \sqrt{3}$.

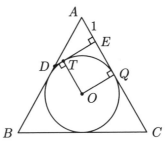

10. (D)   Let $m^2 = n^2 - 440$, where we may assume $m \geq 0$. Then $n^2 - m^2 = 440$, that is,

$$(n{+}m)(n{-}m) = 8 \cdot 5 \cdot 11. \tag{1}$$

Now, n, m are either both even or both odd, or else $n^2 - m^2$ would not be even. Thus $n + m$ and $n - m$ are both even. Therefore, one of them has a factor of 4 and the other a factor of 2, to account for the 8 in (1). There are then four ways to account for the prime factors 5 and 11, that is,

$$\{n{+}m,\ n{-}m\} \ = \ \begin{matrix} \{4{\cdot}5{\cdot}11,\ 2\}, & \{4{\cdot}5,\ 2{\cdot}11\}, \\ \{4{\cdot}11,\ 2{\cdot}5\}, & \text{or } \{4,\ 2{\cdot}5{\cdot}11\}. \end{matrix}$$

In each of these cases, $n - m$ must be the smaller product, since $m \geq 0$. Solving each system, we find 4 integer solutions:

$$(n, m) \ = \ (111, 109),\ (21, 1),\ (27, 17),\ (57, 53).$$

11. (E)   Set P looks like an ice-cream cone lying on its side; see the figures below. The first figure shows how P is obtained. Start with the circle with center $(4,0)$ and radius 2, and shrink it down to the point $(0,0)$ along straight lines. This gives all the points in the circles $C(a)$ with $0 \le a \le 2$. Notice that P fits on and within the two tangent lines from $(0,0)$ to the first circle, $C(2)$. To compute the area of P, dissect it into two 30–60–90 triangles and a circular section of angle $240°$. (See the second figure.) Thus the area is $4\sqrt{3} + \frac{2}{3}4\pi$

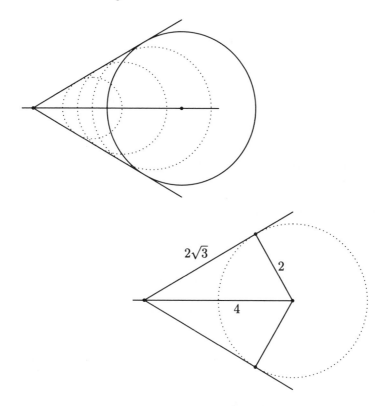

Note. P is the "finite cone" generated by $C(2)$ and the origin. If we allowed infinite stretching away from the origin as well as shrinking to it, we would have the infinite cone generated by $C(2)$. Operations like these on sets are important in certain parts of advanced analysis.

12. (E)   For the first month let A, P, M be, respectively, the amount of sugar the manufacturer could buy (say in kg), the price (\$/kg), and the money (in dollars) the manufacturer had available. Then $A = M/P$. Let A', P', M' be the corresponding values the next month. We seek to express A' as A plus or minus a fraction of A. So

$$A' = \frac{M'}{P'} = \frac{M(1 + \frac{q}{100})}{P(1 + \frac{p}{100})} = A\frac{100 + q}{100 + p} = A\left(1 - \frac{p-q}{100+p}\right).$$

Since $0 < q < p$, then

$$0 < \frac{p - q}{100 + p} < \frac{p - q}{100}.$$

Thus the change from A to A' is a decrease, and it's less than $(p-q)\%$.

Note. If p and q are small (say, less than 10), the decrease is just a little less than $(p-q)\%$. Thus for most problems in the real world, $(p-q)\%$ is a very good estimate. So while (B) is wrong as an exact answer, for practical purposes it is a very good answer. Thus in a certain sense this problem gives the wrong message; that was one reason for deleting it.

13. (A)   First note that when both n and k are positive integers,

$$\binom{n}{k} = \frac{n!}{k!(n-k)!}, \qquad \text{where} \qquad n! = n(n-1)(n-2)\cdots 2 \cdot 1.$$

By the more general definition given in the problem,

$$\binom{\frac{1}{2}}{k} = \frac{(\frac{1}{2})(-\frac{1}{2})\cdots(\frac{1}{2} - k + 1)}{k!} = \frac{(\frac{1}{2})(-\frac{1}{2})\cdots(-\frac{2k-3}{2})}{k!}$$

$$= \frac{\frac{1}{2}(-\frac{1}{2})^{k-1}}{k!}[1 \cdot 3 \cdots (2k-3)]$$

$$= \frac{\frac{1}{2}(-\frac{1}{2})^{k-1}}{k!} \frac{1 \cdot 2 \cdot 3 \cdot 4 \cdots (2k-2)}{2 \cdot 4 \cdots (2k-2)}$$

$$= \frac{\frac{1}{2}(-\frac{1}{2})^{k-1}}{k!} \frac{(2k-2)!}{2^{k-1}(k-1)!}$$

$$= \left(-\frac{1}{4}\right)^{k-1}\frac{1}{2k}\frac{(2k-2)!}{(k-1)!(k-1)!} = \left(-\frac{1}{4}\right)^{k-1}\frac{1}{2k}\binom{2k-2}{k-1},$$

where the last equality follows from the first display in this solution.

14. (D) We give three solutions. The first two use the general Binomial Theorem, which says that if $|b/a| < 1$, then for any *real* power p,

$$(a + b)^p = a^p + pa^{p-1}b + \frac{p(p-1)}{2}a^{p-2}b^2 + \cdots,$$

where the terms continue indefinitely if p is not a nonnegative integer and the right-hand side is interpreted as the limit of a sum. (See any calculus book for further discussion.)

First method. Let $\epsilon_1 = 0.000\,000\,030 = 30 \times 10^{-9}$, and let $\epsilon_2 = 0.000\,000\,015 = 15 \times 10^{-9}$. Now use the Binomial Theorem with $p = 1/3$, $a = 1000$ and $b = \epsilon$, where ϵ will in turn be ϵ_1 and ϵ_2. We get

$$\sqrt[3]{1000 + \epsilon} = 10 + \frac{1}{3}\cdot\frac{1}{100}\epsilon + \text{small multiples of powers of } \epsilon$$
$$\approx 10 + \frac{\epsilon}{300}.$$

Thus $\sqrt[3]{1000 + \epsilon_1} - \sqrt[3]{1000 + \epsilon_2} \approx \frac{\epsilon_1 - \epsilon_2}{300} = 5 \times 10^{-11}$.

Second method. With ϵ_1, ϵ_2 as before, define δ_1, δ_2 so that $10 + \delta_i = \sqrt[3]{1000 + \epsilon_i}$. By the Binomial Theorem with $p = 3$,

$$1000 + \epsilon_i = (10 + \delta_i)^3 = 1000 + 300\delta_i + 30\delta_i^2 + \delta_i^3.$$

Since δ_i is clearly much smaller than 1, a very good approximation of δ_i (many significant digits, and we only need one) is obtained by replacing the right-hand side in the last display by $1000 + 300\delta_i$. Thus $\delta_i \approx \epsilon_i/300$, so

$$\sqrt[3]{1000 + \epsilon_1} + \sqrt[3]{1000 + \epsilon_2} = \delta_1 - \delta_2 \approx \frac{\epsilon_1 - \epsilon_2}{300} = 5 \times 10^{-11}$$

as before.

Third method. We seek $x - y$ where $x = \sqrt[3]{1000 + \epsilon_1}$ and $y = \sqrt[3]{1000 + \epsilon_2}$. Substitute into the factoring identity

$$x - y = \frac{x^3 - y^3}{x^2 + xy + y^2}$$

and approximate the denominator. Since ϵ_i is very small compared to 1000, each term in the denominator is almost exactly 100.

Remark. The message of this problem is: Almost any numerical calculating device (calculator or computer) will have limits on its accuracy, and, when you go beyond those limits, you need mathematics. However, any device that actually allowed you to input the initial data correctly (e.g., $\epsilon_2 = 1000.000\,000\,015$, with 13 significant digits) could presumably compute a sufficiently accurate answer (because to 13 digits, $\sqrt[3]{\epsilon_2} = 10.000\,000\,000\,05$). This is because almost all modern devices use floating point arithmetic, that is, their accuracy does not depend on the location of the decimal point. Most current calculators have 8 or 9 place displays, so you would get 0 as your output because you would have to input both numbers to be cube-rooted as, at best, 1000.00000.

15. (E)   There are not enough books for $n \le 5$. Specifically, the way to give out the smallest number of books consistent with at most $n \le 5$ people getting the same number of books is to divide the 30 people into 6 groups of 5 people each and to give b books to each person in the bth group. But this division requires $5(1 + 2 + 3 + 4 + 5 + 6) = 105$ books. To show that $n = 6$ is possible, it suffices to list a specific division. Let 6 people get b books each for $b = 1, 2, 3, 4$, and let the remaining 6 people get $6, 6, 7, 7, 7, 7$ books respectively.

16. (D)   Let the index of the first base be b. Then the first resident said that $2b + 6$ people use that base and $2b + 2$ use the other, which has index $b + 4$. The second resident must be using the other base, because, expressed with the first base, there are at least 26 people in Abaze and he said there are only 25. So the second resident said that there are $2(b+4) + 5 = 2b + 13$ residents total, that $(b+4) + 3 = b + 7$ use both bases and 1 uses neither. By inclusion-exclusion, the total population equals

$$(2b + 6) + (2b + 2) - (b + 7) + 1 = 3b + 2.$$

Therefore, $2b + 13 = 3b + 2$, so $b = 11$. Thus, using base 10, the population is 35. (28 people use base 11, 24 use base 15, and 18 use both.)

17. (D)   We show that there are two complementary acute angles α satisfying the given equations. Both (A) and (C) are expressions for the smaller angle, (D) is the larger, and (B) is neither. To show all this, we use two approaches to solving the equation. We take greater care than usual to keep track of the location of the angle (it should be in $0 \le \alpha \le \frac{\pi}{2}$).

Approach 1. Square:

$$\sin\alpha + \cos\alpha = \frac{4}{3}, \quad 0 \le \alpha \le \frac{\pi}{2}, \tag{1}$$

$$1 + 2\sin\alpha\cos\alpha = \frac{16}{9}, \quad 0 \le \alpha \le \frac{\pi}{2}, \tag{2}$$

$$\sin 2\alpha = \frac{7}{9}, \quad 0 \le 2\alpha \le \pi. \tag{3}$$

Lines (1) and (3) are equivalent: there is no problem about getting $\pm\frac{4}{3}$ when reversing from (2) to (1) because the bounds on α force $\sin\alpha + \cos\alpha$ to be positive. In (3), 2α goes beyond the range of the function $\sin^{-1}$. Thus (3) has two supplementary solutions for 2α:

$$2\alpha = \sin^{-1}\frac{7}{9}, \qquad \text{(smaller)}$$

$$2\alpha = \pi - \sin^{-1}\frac{7}{9}. \qquad \text{(larger)}$$

Thus the larger α is not (A) but rather $\frac{1}{2}\left(\pi - \sin^{-1}\frac{7}{9}\right)$. Unfortunately, this is not among the answers given.

To test whether (B) is an equivalent expression for the correct answer, note that (3) implies $\cos 2\alpha = \pm\frac{4\sqrt{2}}{9}$, the sign depending on whether 2α is in quadrant I or II. Thus for the larger acute α, 2α is in quadrant II. That is, $\alpha = \frac{1}{2}\cos^{-1}\left(-\frac{4\sqrt{2}}{9}\right)$. This is not the same angle as (B), since $\cos^{-1}$ is a 1-to-1 function.

Approach 2. Multiply the given equation by $\sin\frac{\pi}{4} = \cos\frac{\pi}{4} = \frac{\sqrt{2}}{2}$. One gets the equivalent equations

$$\sin\alpha\sin\frac{\pi}{4} + \cos\alpha\cos\frac{\pi}{4} = \frac{4}{3}\frac{\sqrt{2}}{2} = \frac{2\sqrt{2}}{3}, \quad 0 \le \alpha \le \frac{\pi}{2},$$

$$\cos\left(\alpha - \frac{\pi}{4}\right) = \frac{2\sqrt{2}}{3}, \quad -\frac{\pi}{4} \le \alpha - \frac{\pi}{4} \le \frac{\pi}{4}. \tag{4}$$

Since the range of $\cos^{-1}$ is $[0, \pi]$, there are two values of $\alpha - \frac{\pi}{4}$ satisfying both parts of (4), namely $\pm\cos^{-1}\left(\frac{2\sqrt{2}}{3}\right)$. Thus (D) is the larger acute value for α and

$$\frac{\pi}{4} - \cos^{-1}\left(\frac{2\sqrt{2}}{3}\right) = \frac{\pi}{4} - \left(\frac{\pi}{2} - \sin^{-1}\left(\frac{2\sqrt{2}}{3}\right)\right) = \sin^{-1}\left(\frac{2\sqrt{2}}{3}\right) - \frac{\pi}{4}$$

is the smaller angle. This is answer (C).

Two correct expressions for the larger α were derived above. There are many other correct expressions that could have been used instead of (D), and many other expressions for the smaller α that could have been further distractors! Can you find a number c such that $\alpha = \sin^{-1} c$?

18. (D) $a^3 + b^3 = (a+b)(a^2-ab+b^2)$, and we will know both factors on the right as soon as we know ab. Now, $a + b = 1$ implies $a^2 + 2ab + b^2 = 1$, so subtracting $a^2 + b^2 = 2$ from this square gives $2ab = -1$. Thus

$$a^3 + b^3 = (a+b)[(a^2+b^2) - ab] = 1(2 + \tfrac{1}{2}) = 2\tfrac{1}{2}.$$

19. (C)   For a circle of radius r to be tangent to the parabola at $x = 0$, it must have a horizontal tangent there, so its center must be at $(0, r)$. Moreover, this circles lies on or inside the parabola iff, for all real x, the point (x, x^2) on the parabola has distance at least r from $(0, r)$. This is equivalent to the following statements:

$$(x - 0)^2 + (x^2 - r)^2 \geq r^2 \quad \text{for all } x$$
$$x^2 + (x^2 - r)^2 - r^2 \geq 0 \quad \text{for all } x$$
$$x^2\left(x^2 + (1-2r)\right) \geq 0 \quad \text{for all } x$$
$$x^2 + (1-2r) \geq 0 \quad \text{for all } x$$
$$1 - 2r \geq 0$$
$$1/2 \geq r.$$

Thus there is a largest radius, $1/2$.

Equivalently, one can find the largest r for which the simultaneous equations

$$y = x^2, \qquad (x-0)^2 + (y-r)^2 = r^2$$

have a unique real solution $(x, y) = (0, 0)$.

Note. The radius of this circle happens to be the "radius of curvature" of the parabola at $x = 0$, so students who have learned how to compute radius of curvature in calculus can find r this way. (This is not an easier approach unless you happen to remember the formula.) But this approach works only because the apex of a parabola is the point at which it curves the fastest. The circle of largest radius that is inside $y = x^2$ and tangent at, say, $(\frac{1}{2}, \frac{1}{4})$ is *not* the circle with the radius of curvature there. Can you find the radius in this case?

20. (E)   First note that $\log_a b$ is defined iff $[a > 0,\ a \neq 1$ and $b > 0]$. Thus $\log_x(x+1)$ is defined iff $[x > 0,\ x \neq 1$ and $x+1 > 0]$, i.e., iff $[x > 0$ and $x \neq 1]$. This eliminates choices (A) and (B). Now let $\log x$ mean $\log_{10} x$. Then

$$\begin{aligned}
\log_x(x+1) &= \frac{\log(x+1)}{\log x} \\
&= \frac{\log x + \log\left(\frac{x+1}{x}\right)}{\log x} \\
&= 1 + \frac{\log\left(1 + \frac{1}{x}\right)}{\log x}.
\end{aligned}$$

When $0 < x < 1$, we have $\log x < 0$, $\log(x+1) > 0$. Thus from the first line of the display above, $\log_x(x+1) < 0$ for such x. This eliminates (C). (In fact, $\lim\limits_{x \to 0} \dfrac{\log(x+1)}{\log x} = \dfrac{0}{-\infty} = 0$, which explains the behavior of the graph near $x = 0$ in the remaining candidates, (D) and (E), but we won't need this fact or other knowledge of limits to eliminate one of these choices.) To eliminate (D), use the third line of the display to show that $\log_x(x+1)$ is a decreasing function for x increasing from 1. In detail, $1 + \frac{1}{x}$ decreases, so $\log(1+\frac{1}{x})$ decreases; since $\log x$ increases, the third line decreases. (In fact, $\lim\limits_{x \to \infty} \log_x(x+1) = 1$.)

21. (D)   For any $a > 0$, $\log_a x$ is defined only for $x > 0$ (except for $a = 1$, where $\log_a x$ is not defined for any x). Thus for $\log_{1/2}\left(\log_2(\log_{1/2} x)\right)$ to be defined, three inequalities must be satisfied:

   (i) $x > 0$;

   (ii) $\log_{1/2} x > 0$, which is equivalent to $0 < x < 1$;

   (iii) $\log_2(\log_{1/2} x) > 0 \iff \log_{1/2} x > 1 \iff 0 < x < 1/2$.

These three conditions are satisfied if and only if $0 < x < 1/2$.

22. (C)   Suppose that the arm lengths of the balance are L and $1.05L$. Without loss of generality, we may assume pan A is at the end of the arm of length L. For the first weighing, we then have $210L = W_1(1.05L)$, or $W_1 = 210/1.05 = 200$ grams. For the second weighing, $W_2 L = 210(1.05)L$, or $W_2 = 220.5$ grams. Therefore, $W_1 + W_2 = 220 + 220.5 = 420.5$ grams.

Query. Can you show that, no matter by what percent p one arm is longer than the other, this symmetrization method will result in too much material being measured, but always by $< \frac{1}{2}p$ percent (usually much less).

23. (B)   Let s be the common side length of the triangle. Also, let $a = AF$, $b = BF$, $m = AD$ and $n = BE$. By the theorem relating secants to tangents,

$$ms = a^2,$$
$$ns = b^2.$$

Dividing yields

$$\frac{m}{n} = \frac{a^2}{b^2} = p^2.$$

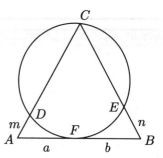

Query. Need $\triangle ABC$ be equilateral?

24. (D) Since $0 < r < 1$, we have $0 < 11r < 11$, so $\lfloor 11r \rfloor$ is $0, 1, \ldots$ or 10. Moreover, since $11r$ is equally likely to be in any one of the intervals

$$(0,1), \quad [1,2), \quad [2,3), \quad \ldots \quad [10,11),$$

the integers $0, 1, \ldots, 10$ are equally likely output. (The interval $(0,1)$ contains neither endpoint, while the others contain left endpoints. Nonetheless, $11r$ is equally likely to be in any of these intervals because the probability is 0 that $11r$ will equal an integer exactly.)

Note. Answers (B) and (E) always give 0. (A) gives a random integer from 0 to 9. (C) gives an integer from 0 to 10, but they are not all equally likely. To see this, observe that $.5 < 10r + .5 < 10.5$. Therefore, $\lfloor 10r + .5 \rfloor$ equals 0 if $10r + .5$ is in the interval $(.5, 1)$ and equals 1 if $10r + .5$ is in the interval $[1, 2)$. Since the latter interval is twice as long, 1 is twice as likely an output as 0.

25. (C) Let P be the center of the circle and a be the side length of the square. Then on the basis of the adjoining figure and the Pythagorean Theorem, $r^2 = \left(\frac{a}{2}\right)^2 + (a-r)^2$. Upon expansion and simplification, this equation yields $a = \frac{8}{5}r$. Hence the area of the square is $\frac{64}{25}r^2$.

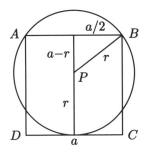

26. (E) We want 95% probability that the total number of heads will be within 10% of 50%, i.e., within $n/10$ of $n/2$. From the fact stated in the problem, we must pick n so that $\sqrt{n} = n/10$. Solving gives $n = 100$.

Where does this fact come from? Coin tossing is an example of a *Binomial Distribution*, used for studying the number of successes in n trials when each trial has probability of success p. In our case, "success" is a head and $p = \frac{1}{2}$. It can be shown that, as n gets large, the Binomial Distribution approaches the so-called *Normal Distribution* with mean np and standard deviation $\sqrt{np(1-p)}$. Thus one may approximate the probabilities for the various Binomial Distributions using the Normal Distribution. In our case, $np = \frac{1}{2}n$ and $\sqrt{np(1-p)} = \frac{1}{2}\sqrt{n}$. Thus the question

What is the probability that the number of heads will be within $\sqrt{n}$ of $n/2$?

becomes

What is the probability that a random value taken from the Normal Distribution is within 2 standard deviations of the mean?

The answer, taken from a table, is 95.4%. For a full discussion, see any probability text which assumes calculus, e.g., W. Feller, "Introduction to Probability Theory and Its Applications, Vol. 1," 3rd ed., Wiley, New York, 1968.

Observation. It takes more coin tosses than most people think before the fraction of heads is very likely to be very close to $\frac{1}{2}$.

27. (D)   To find a factorization of

$$A = ab(c^2 + d^2) + cd(a^2 + b^2) = abc^2 + abd^2 + a^2cd + b^2cd,$$

we try regrouping the terms. For instance,

$$A = (abc^2 + a^2cd) + (abd^2 + b^2cd)$$
$$= ac(bc + ad) + bd(ad + bc)$$
$$= (ac + bd)(bc + ad).$$

Note. The values $a = 1$, $b = 3$, $c = 2$, $d = 5$ show that choices (A), (B) and (C) are all incorrect; why?

Also, can you argue the incorrectness of these choices algebraically rather than by plug-in?

1983 AIME

1. Let x, y and z all exceed 1 and let w be a positive number such that
$$\log_x w = 24, \ \log_y w = 40 \ \text{and} \ \log_{xyz} w = 12.$$
Find $\log_z w$.

2. Let $f(x) = |x-p| + |x-15| + |x-p-15|$, where $0 < p < 15$. Determine the minimum value taken by $f(x)$ for x in the interval $p \le x \le 15$.

3. What is the product of the real roots of the equation
$$x^2 + 18x + 30 = 2\sqrt{x^2 + 18x + 45}?$$

4. A machine-shop cutting tool has the shape of a notched circle, as shown. The radius of the circle is $\sqrt{50}$ cm, the length of AB is 6 cm and that of BC is 2 cm. The angle ABC is a right angle. Find the square of the distance (in centimeters) from B to the center of the circle.

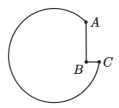

5. Suppose that the sum of the squares of two complex numbers x and y is 7 and the sum of the cubes is 10. What is the largest real value that $x+y$ can have?

6. Let $a_n = 6^n + 8^n$. Determine the remainder on dividing a_{83} by 49.

7. Twenty five of King Arthur's knights are seated at their customary round table. Three of them are chosen – all choices of three being equally likely – and are sent off to slay a troublesome dragon. Let P be the probability that at least two of the three had been sitting next to each other. If P is written as a fraction in lowest terms, what is the sum of the numerator and denominator?

8. What is the largest 2-digit prime factor of the integer $n = \binom{200}{100}$?

9. Find the minimum value of
$$f(x) = \frac{9x^2 \sin^2 x + 4}{x \sin x}$$
for $0 < x < \pi$.

10. The numbers 1447, 1005 and 1231 have something in common: each is a 4-digit number beginning with 1 that has exactly two identical digits. How many such numbers are there?

11. The solid shown has a square base of side length s. The upper edge is parallel to the base and has length $2s$. All other edges have length s. Given that $s = 6\sqrt{2}$, what is the volume of the solid?

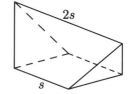

12. Diameter AB of a circle has length a 2-digit integer (base ten). Reversing the digits gives the length of the perpendicular chord CD. The distance from their intersection point H to the center O is a positive rational number. Determine the length of AB.

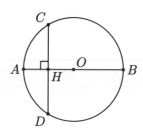

13. For $\{1, 2, 3, \ldots, n\}$ and each of its nonempty subsets a unique **alternating sum** is defined as follows: Arrange the number in the subset in decreasing order and then, beginning with the largest, alternately add and subtract successive numbers. (For example, the alternating sum for $\{1, 2, 3, 6, 9\}$ is $9-6+4-2+1 = 6$ and for $\{5\}$ it is simply 5.) Find the sum of all such alternating sums for $n = 7$.

14. In the adjoining figure, two circles of radii 8 and 6 are drawn with their centers 12 units apart. At P, one of the points of intersection, a line is drawn in such a way that the chords QP and PR have equal length. Find the square of the length of QP.

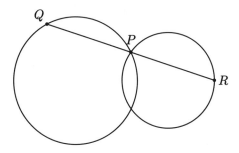

15. The adjoining figure shows two intersecting chords in a circle, with B on minor arc AD. Suppose that the radius of the circle is 5, that $BC = 6$, and that AD is bisected by BC. Suppose further that AD is the only chord starting at A which is bisected by BC. It follows that the sine of the central angle of minor arc AB is a rational number. If this number is expressed as a fraction m/n in lowest terms, what is the product mn?

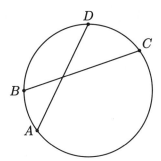

1984 AIME

1. Find the value of $a_2 + a_4 + a_6 + \cdots + a_{98}$ if $a_1, a_2, a_3, \ldots$ is an arithmetic progression with common difference 1, and $a_1 + a_2 + a_3 + \cdots + a_{98} = 137$.

2. The integer n is the smallest positive multiple of 15 such that every digit of n is either 0 or 8. Compute $\frac{n}{15}$.

3. A point P is chosen in the interior of $\triangle ABC$ so that when lines are drawn through P parallel to the sides of $\triangle ABC$, the resulting smaller triangles, t_1, t_2 and t_3 in the figure, have areas 4, 9 and 49, respectively. Find the area of $\triangle ABC$.

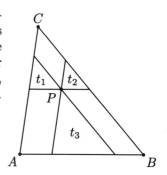

4. Let S be a list of positive integers – not necessarily distinct – in which the number 68 appears. The average (arithmetic mean) of the numbers in S is 56. However, if 68 is removed, the average of the remaining numbers drops to 55. What is the largest number that can appear in S?

5. Determine the value of ab if
$$\log_8 a + \log_4 b^2 = 5 \quad \text{and} \quad \log_8 b + \log_4 a^2 = 7.$$

6. Three circles, each of radius 3, are drawn with centers at (14,92), (17,76) and (19,84). A line passing through (17,76) is such that the total area of the parts of the three circles to one side of the line is equal to the total area of the parts of the three circles to the other side of it. What is the absolute value of the slope of this line?

7. The function f is defined on the set of integers and satisfies

$$f(n) = \begin{cases} n - 3, & \text{if } n \geq 1000. \\ f(f(n+5)), & \text{if } n < 1000. \end{cases}$$

Find $f(84)$.

8. The equation $z^6 + z^3 + 1 = 0$ has one complex root with argument (angle) θ between $90°$ and $180°$ in the complex plane. Determine the degree measure of θ.

9. In tetrahedron $ABCD$, edge AB has length 3 cm. The area of face ABC is 15 cm^2 and the area of face ABD is 12 cm^2. These two faces meet each other at a $30°$ angle. Find the volume of the tetrahedron in cm^3.

10. Mary told John her score on the American High School Mathematics Examination (AHSME), which was over 80. From this, John was able to determine the number of problems Mary solved correctly. If Mary's score had been any lower, but still over 80, John could not have determined this. What was Mary's score? (Recall that the AHSME consists of 30 multiple-choice problems and that one's score, s, is computed by the formula $s = 30 + 4c - w$, where c is the number of correct and w is the number of wrong answers; students are not penalized for problems left unanswered.)

11. A gardener plants three maple trees, four oak trees and five birch trees in a row. He plants them in random order, each arrangement being equally likely. Let $\frac{m}{n}$ in lowest terms be the probability that no two birch trees are next to one another. Find $m + n$.

12. A function f is defined for all real numbers and satisfies

$$f(2+x) = f(2-x) \quad \text{and} \quad f(7+x) = f(7-x)$$

for all real x. If $x = 0$ is a root of $f(x) = 0$, what is the least number of roots $f(x) = 0$ must have in the interval $-1000 \leq x \leq 1000$?

13. Find the value of $10\cot(\cot^{-1} 3 + \cot^{-1} 7 + \cot^{-1} 13 + \cot^{-1} 21)$.

14. What is the largest even integer which cannot be written as the sum of two odd composite numbers? (Recall that a positive integer is said to be composite if it is divisible by at least one positive integer other than 1 and itself.)

15. Determine $x^2 + y^2 + z^2 + w^2$ if

$$\frac{x^2}{2^2 - 1^2} + \frac{y^2}{2^2 - 3^2} + \frac{z^2}{2^2 - 5^2} + \frac{w^2}{2^2 - 7^2} = 1,$$

$$\frac{x^2}{4^2 - 1^2} + \frac{y^2}{4^2 - 3^2} + \frac{z^2}{4^2 - 5^2} + \frac{w^2}{4^2 - 7^2} = 1,$$

$$\frac{x^2}{6^2 - 1^2} + \frac{y^2}{6^2 - 3^2} + \frac{z^2}{6^2 - 5^2} + \frac{w^2}{6^2 - 7^2} = 1,$$

$$\frac{x^2}{8^2 - 1^2} + \frac{y^2}{8^2 - 3^2} + \frac{z^2}{8^2 - 5^2} + \frac{w^2}{8^2 - 7^2} = 1.$$

1985 AIME

1. Let $x_1 = 97$, and for $n > 1$ let $x_n = \frac{n}{x_{n-1}}$. Calculate the product $x_1 x_2 \ldots x_8$.

2. When a right triangle is rotated about one leg, the volume of the cone produced is 800π cm^3. When the triangle is rotated about the other leg, the volume of the cone produced is 1920π cm^3. What is the length (in cm) of the hypotenuse of the triangle?

3. Find c if a, b and c are positive integers which satisfy $c = (a + bi)^3 - 107i$, where $i^2 = -1$.

4. A small square is constructed inside a square of area 1 by dividing each side of the unit square into n equal parts, and then connecting the vertices to the division points closest to the opposite vertices, as shown in the figure on the right. Find the value of n if the area of the small square (shaded in the figure) is exactly 1/1985.

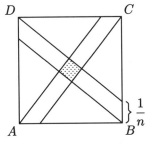

5. A sequence of integers $a_1, a_2, a_3, \ldots$ is chosen so that $a_n = a_{n-1} - a_{n-2}$ for each $n \geq 3$. What is the sum of the first 2001 terms of this sequence if the sum of the first 1492 terms is 1985, and the sum of the first 1985 terms is 1492?

6. As shown in the figure on the right, $\triangle ABC$ is divided into six smaller triangles by lines drawn from the vertices through a common interior point. The areas of four of these triangles are as indicated. Find the area of $\triangle ABC$.

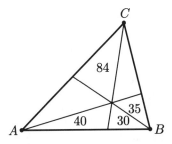

7. Assume that a, b, c and d are positive integers such that $a^5 = b^4$, $c^3 = d^2$ and $c - a = 19$. Determine $d - b$.

163

8. The sum of the following seven numbers is exactly 19:

$$a_1 = 2.56, \qquad a_2 = 2.61, \qquad a_3 = 2.65, \qquad a_4 = 2.71,$$

$$a_5 = 2.79, \qquad a_6 = 2.82, \qquad a_7 = 2.86.$$

It is desired to replace each a_i by an integer approximation A_i, $1 \le i \le 7$, so that the sum of the A_i's is also 19, and so that M, the maximum of the "errors" $|A_i - a_i|$, is as small as possible. For this minimum M, what is $100M$?

9. In a circle, parallel chords of lengths 2, 3 and 4 determine central angles of α, β and $\alpha + \beta$ radians, respectively, where $\alpha + \beta < \pi$. If $\cos\alpha$, which is a positive rational number, is expressed as a fraction in lowest terms, what is the sum of its numerator and denominator?

10. How many of the first 1000 positive integers can be expressed in the form

$$\lfloor 2x \rfloor + \lfloor 4x \rfloor + \lfloor 6x \rfloor + \lfloor 8x \rfloor,$$

where x is a real number, and $\lfloor z \rfloor$ denotes the greatest integer less than or equal to z?

11. An ellipse has foci at (9,20) and (49,55) in the xy-plane and is tangent to the x−axis. What is the length of its major axis?

12. Let A, B, C and D be the vertices of a regular tetrahedron, each of whose edges measures 1 meter. A bug, starting from vertex A, observes the following rule: at each vertex it chooses one of the three edges meeting at that vertex, each edge being equally likely to be chosen, and crawls along that edge to the vertex at its opposite end. Let $p = n/729$ be the probability that the bug is at vertex A when it has crawled exactly 7 meters. Find the value of n.

13. The numbers in the sequence 101, 104, 109, 116,... are of the form $a_n = 100 + n^2$, where $n = 1, 2, 3, \ldots$. For each n, let d_n be the greatest common divisor of a_n and a_{n+1}. Find the maximum value of d_n as n ranges through the positive integers.

14. In a tournament each player played exactly one game against each of the other players. In each game the winner was awarded 1 point, the loser got 0 points, and each of the two players earned 1/2 point if the game was a tie. After the completion of the tournament, it was found that exactly half of the points earned by each player were earned in games against the ten players with the least number of points. (In particular, each of the ten lowest scoring players earned half of her/his points against the other nine of the ten). What was the total number of players in the tournament?

15. Three 12 cm × 12 cm squares are each cut into two pieces A and B, as shown in the first figure below, by joining the midpoints of two adjacent sides. These six pieces are then attached to a regular hexagon, as shown in the second figure, so as to fold into a polyhedron. What is the volume (in cm^3) of this polyhedron?

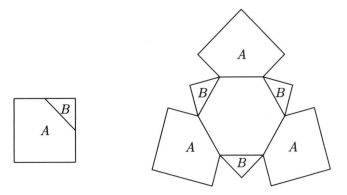

1986 AIME

1. What is the sum of the solutions of the equation

$$\sqrt[4]{x} = \frac{12}{7 - \sqrt[4]{x}}?$$

2. Evaluate the product

$$(\sqrt{5} + \sqrt{6} + \sqrt{7})(\sqrt{5} + \sqrt{6} - \sqrt{7})(\sqrt{5} - \sqrt{6} + \sqrt{7})(-\sqrt{5} + \sqrt{6} + \sqrt{7}).$$

3. If $\tan x + \tan y = 25$ and $\cot x + \cot y = 30$, what is $\tan(x + y)$?

4. Determine $3x_4 + 2x_5$, if x_1, x_2, x_3, x_4 and x_5 satisfy the system of equations given below:

$$
\begin{aligned}
2x_1 + x_2 + x_3 + x_4 + x_5 &= 6 \\
x_1 + 2x_2 + x_3 + x_4 + x_5 &= 12 \\
x_1 + x_2 + 2x_3 + x_4 + x_5 &= 24 \\
x_1 + x_2 + x_3 + 2x_4 + x_5 &= 48 \\
x_1 + x_2 + x_3 + x_4 + 2x_5 &= 96.
\end{aligned}
$$

5. What is the largest positive integer n for which $n^3 + 100$ is divisible by $n + 10$?

6. The pages of a book are numbered 1 though n. When the page numbers of the book were added, one of the page numbers was mistakenly added twice, resulting in the incorrect sum of 1986. What was the number of the page that was added twice?

7. The increasing sequence 1, 3, 4, 9, 10, 12, 13,... consists of all those positive integers which are powers of 3 or sums of distinct powers of 3. Find the 100th term of this sequence (where 1 is the 1st term, 3 is the 2nd term, and so on).

8. Let S be the sum of the base 10 logarithms of all of the proper divisors of 1,000,000. (By a proper divisor of a natural number we mean a positive integral divisor other than 1 and the number itself.) What is the integer nearest to S?

166

9. In $\triangle ABC$ shown below, $AB = 425$, $BC = 450$ and $CA = 510$. Moreover, P is an interior point chosen so that the segments DE, FG and HI are each of length d, contain P, and are parallel to the sides AB, BC and CA, respectively. Find d.

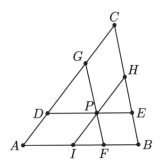

10. In a parlor game the "magician" asks one of the participants to think of a three-digit number (abc), where a, b and c represent digits in base 10 in the order indicated. Then the magician asks this person to form the numbers (acb), (bac), (bca), (cab) and (cba), to add these five numbers, and to reveal their sum, N. If told the value of N, the magician can identify the original number, (abc). Play the role of the magician and determine (abc) if $N = 3194$.

11. The polynomial $1 - x + x^2 - x^3 + \cdots + x^{16} - x^{17}$ may be written in the form $a_0 + a_1 y + a_2 y^2 + a_3 y^3 + \cdots + a_{16} y^{16} + a_{17} y^{17}$, where $y = x + 1$ and the a_i's are constants. Find the value of a_2.

12. Let the sum of a set of numbers be the sum of its elements. Let S be a set of positive integers, none greater than 15. Suppose no two disjoint subsets of S have the same sum. What is the largest sum a set S with these properties can have?

13. In a sequence of coin tosses one can keep a record of the number of instances when a tail is immediately followed by a head, a head is immediately followed by a head, etc. We denote these by TH, HH, etc. For example, in the sequence HHTTHHHHTHHTTTT of 15 coin tosses we observe that there are five HH, three HT, two TH and four TT subsequences. How many different sequences of 15 coin tosses will contain exactly two HH, three HT, four TH and five TT subsequences?

14. The shortest distances between an interior diagonal of a rectangular parallelepiped (box), P, and the edges it does not meet are $2\sqrt{5}$, $30/\sqrt{13}$ and $15/\sqrt{10}$. Determine the volume of P.

15. Let $\triangle ABC$ be a right triangle in the xy-plane with the right angle at C. Given that the length of the hypotenuse AB is 60, and that the medians through A and B lie along the lines $y = x + 3$ and $y = 2x + 4$, respectively, find the area of $\triangle ABC$.

1987 AIME

1. An ordered pair (m, n) of non-negative integers is called "simple" if the addition $m + n$ in base 10 requires no carrying. Find the number of simple ordered pairs of non-negative integers that sum to 1492.

2. What is the largest possible distance between two points, one on the sphere of radius 19 with center $(-2, -10, 5)$ and the other on the sphere of radius 87 with center $(12, 8, -16)$?

3. By a proper divisor of a natural number we mean a positive integral divisor other than 1 and the number itself. A natural number greater than 1 will be called "nice" if it is equal to the product of its distinct proper divisors. What is the sum of the first ten nice numbers?

4. Find the area of the region enclosed by the graph of $|x-60|+|y| = |x/4|$.

5. Find $3x^2y^2$ if x and y are integers such that $y^2+3x^2y^2 = 30x^2+517$.

6. Rectangle $ABCD$ is divided into four parts of equal area by five segments as shown in the figure, where $XY = YB + BC + CZ = ZW = WD + DA + AX$, and PQ is parallel to AB. Find the length of AB (in cm) if $BC = 19$ cm and $PQ = 87$ cm.

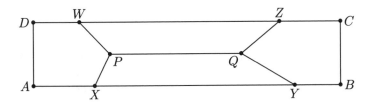

7. Let $[r, s]$ denote the least common multiple of positive integers r and s. Find the number of ordered triples (a, b, c) of positive integers for which $[a, b] = 1000$, $[b, c] = 2000$, and $[c, a] = 2000$.

8. What is the largest positive integer n for which there is a unique integer k such that $\frac{8}{15} < \frac{n}{n+k} < \frac{7}{13}$?

9. Triangle ABC has right angle at B, and contains a point P for which $PA = 10, PB = 6$ and $\angle APB = \angle BPC = \angle CPA$. Find PC.

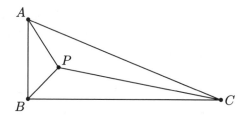

10. Al walks down to the bottom of an escalator that is moving up and he counts 150 steps. His friend, Bob, walks up to the top of the escalator and counts 75 steps. If Al's speed of walking (in steps per unit time) is three times Bob's speed, how many steps are visible on the escalator at any given time? (Assume that this number is constant.)

11. Find the largest possible value of k for which 3^{11} is expressible as the sum of k consecutive positive integers.

12. Let m be the smallest positive integer whose cube root is of the form $n + r$, where n is a positive integer and r is a positive real number less than $1/1000$. Find n.

13. A given sequence $r_1, r_2, \ldots, r_n$ of distinct real numbers can be put in ascending order by means of one or more "bubble passes". A bubble pass through a given sequence consists of comparing the second term with the first term and exchanging them if and only if the second term is smaller, then comparing the third term with the current second term and exchanging them if and only if the third term is smaller, and so on in order, through comparing the last term, r_n, with its current predecessor and exchanging them if and only if the last term is smaller.

The example on the right shows how the sequence 1, 9, 8, 7 is transformed into the sequence 1, 8, 7, 9 by one bubble pass. The numbers compared at each step are underlined.

$$
\begin{array}{cccc}
\underline{1} & \underline{9} & 8 & 7 \\
1 & \underline{9} & \underline{8} & 7 \\
1 & 8 & \underline{9} & \underline{7} \\
1 & 8 & 7 & 9
\end{array}
$$

Suppose that $n = 40$, and that the terms of the initial sequence $r_1, r_2, \ldots, r_{40}$ are distinct from one another and are in random order. Let p/q, in lowest terms, be the probability that the number that begins as r_{20} will end up, after one bubble pass, in the 30th place (i.e., will have 29 terms on its left and 10 terms on its right). Find $p + q$.

14. Compute

$$
\frac{(10^4 + 324)(22^4 + 324)(34^4 + 324)(46^4 + 324)(58^4 + 324)}{(4^4 + 324)(16^4 + 324)(28^4 + 324)(40^4 + 324)(52^4 + 324)}.
$$

15. Squares S_1 and S_2 are inscribed in right triangle ABC, as shown in the figures below. Find $AC + CB$ if area$(S_1) = 441$ and area$(S_2) = 440$.

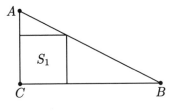
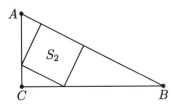

1988 AIME

1. One commercially available ten-button lock may be opened by depressing — in any order — the correct five buttons. The sample shown at right has $\{1, 2, 3, 6, 9\}$ as its combination. Suppose that these locks are redesigned so that sets of as many as nine buttons or as few as one button could serve as combinations. How many additional combinations would this allow?

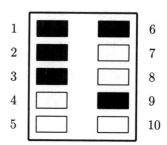

2. For any positive integer k, let $f_1(k)$ denote the square of the sum of the digits of k. For $n \geq 2$, let $f_n(k) = f_1(f_{n-1}(k))$. Find $f_{1988}(11)$.

3. Find $(\log_2 x)^2$ if $\log_2(\log_8 x) = \log_8(\log_2 x)$.

4. Suppose that $|x_i| < 1$ for $i = 1, 2, \ldots, n$. Suppose further that
$$|x_1| + |x_2| + \ldots + |x_n| = 19 + |x_1 + x_2 + \ldots + x_n|.$$
What is the smallest possible value of n?

5. Let m/n, in lowest terms, be the probability that a randomly chosen positive divisor of 10^{99} is an integer multiple of 10^{88}. Find $m + n$.

6. It is possible to place positive integers into the vacant twenty-one squares of the 5×5 square shown on the right so that the numbers in each row and column form arithmetic sequences. Find the number that must occupy the vacant square marked by the asterisk ($*$).

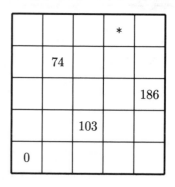

7. In $\triangle ABC$, $\tan(\angle CAB) = 22/7$ and the altitude from A divides BC into segments of length 3 and 17. What is the area of $\triangle ABC$?

8. The function f, defined on the set of ordered pairs of positive integers, satisfies the following properties:

$$f(x,x) = x, \ f(x,y) = f(y,x), \ \text{and} \ (x+y)f(x,y) = yf(x,x+y).$$

Calculate $f(14,52)$.

9. Find the smallest positive integer whose cube ends in 888.

10. A convex polyhedron has for its faces 12 squares, 8 regular hexagons, and 6 regular octagons. At each vertex of the polyhedron one square, one hexagon, and one octagon meet. How many segments joining vertices of the polyhedron lie in the interior of the polyhedron rather than along an edge or a face?

11. Let $w_1, w_2, \ldots, w_n$ be complex numbers. A line L in the complex plane is called a *mean line* for the points $w_1, w_2, \ldots, w_n$ if L contains points (complex numbers) $z_1, z_2, \ldots, z_n$ such that

$$\sum_{k=1}^{n} (z_k - w_k) = 0.$$

For the numbers $w_1 = 32 + 170i$, $w_2 = -7 + 64i$, $w_3 = -9 + 200i$, $w_4 = 1 + 27i$, and $w_5 = -14 + 43i$ there is a unique mean line with y-intercept 3. Find the slope of this mean line.

12. Let P be an interior point of $\triangle ABC$ and extend lines from the vertices through P to the opposite sides. Let a, b, c, and d denote the lengths of the segments indicated in the figure. Find the product abc if $a+b+c = 43$ and $d = 3$.

13. Find a if a and b are integers such that $x^2 - x - 1$ is a factor of $ax^{17} + bx^{16} + 1$.

14. Let C be the graph of $xy = 1$, and denote by C^* the reflection of C in the line $y = 2x$. Let the equation of C^* be written in the form

$$12x^2 + bxy + cy^2 + d = 0.$$

Find the product bc.

15. In an office at various times during the day, the boss gives the secretary a letter to type, each time putting the letter on top of the pile in the secretary's in-box. When there is time, the secretary takes the top letter off the pile and types it. There are nine letters to be typed during the day, and the boss delivers them in the order 1, 2, 3, 4, 5, 6, 7, 8, 9.

While leaving for lunch, the secretary tells a colleague that letter 8 has already been typed, but says nothing else about the morning's typing. The colleague wonders which of the nine letters remain to be typed after lunch and in what order they will be typed. Based upon the above information, how many such *after-lunch typing orders* are possible? (That there are no letters left to be typed is one of the possibilities.)

Dropped AIME Problems

1. Determine the number of five-digit integers $(37abc)$ in base 10 such that each of the numbers $(37abc)$, $(37bca)$ and $(37cab)$ is divisible by 37.

2. Determine the value of $10(x + y)$, given that

$$\lfloor x \rfloor + \lfloor y \rfloor + y = 43.8$$

and

$$x + y - \lfloor x \rfloor = 18.4.$$

(Note: $\lfloor z \rfloor$ denotes the greatest integer less than or equal to z)

3. In the 3×3 array shown on the right, each letter stands for a distinct non-zero integer. Each of the three-digit numbers (abc), (def), (ghi), (adg), (beh), (cfi) and (aei) is divisible by 11. What is the maximum value of the three-digit number (ceg)? (A three-digit number (xyz) is to be interpreted as $100x + 10y + z$.)

$$
\begin{array}{ccc}
a & b & c \\
d & e & f \\
g & h & i
\end{array}
$$

4. In $\triangle ABC$ let D and E be points on sides AC and BC, respectively, so that DE is parallel to AB. Let P be the intersection of AE and BD, and assume that area$(\triangle ABP) = 36$ and area$(\triangle EDP) = 25$. Find area$(\triangle ABC)$.

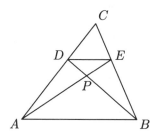

5. Let x and y be real numbers which differ in absolute value and satisfy the equations

$$
\begin{aligned}
x^3 &= 13x + 3y \\
y^3 &= 3x + 13y.
\end{aligned}
$$

Find $(x^2 - y^2)^2$.

6. For each arrangement of the ten numbers 1, 4, 7, 10, 13, 16, 19, 22, 25 and 28 around a circle, let N denote the largest of the ten sums obtained by adding each number to the sum of its neighbors. What is the smallest value of N which can be obtained?

7. Find $x^2 - yz$ if x, y and z are distinct positive integers such that $y^2 - zx = -103$ and $z^2 - xy = 22$.

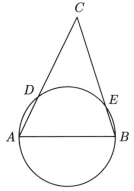

8. In the figure, ABC is a triangle and $AB = 30$ is a diameter of the circle. If $AD = AC/3$ and $BE = BC/4$, then what is the area of the triangle?

9. Define a positive integer N to have the reversed-sum property if there is a three–digit number (abc) with neither a nor c equal to zero and such that $(abc) + (cba) = N$. How many positive integers have the reversed-sum property?

10. Point P is located inside $\triangle ABC$ so that $\angle PAB = \angle PBC = \angle PCA$. The sides of the triangle are $AB = 13$, $BC = 14$, and $CA = 15$. The tangent of $\angle PAB$ may be expressed in the form m/n, where m and n are relatively prime positive integers. Find $m + n$.

11. Let us call a positive integer n "fortunate" if the sum of its digits is a multiple of 7, and "superfortunate" if it is fortunate and if none of the numbers $n + 1, n + 2, \ldots, n + 12$ is fortunate. Find the smallest superfortunate number.

12. Let S be a subset of $\{1, 2, 3, \ldots, 1000\}$ so that, if a and b are any (not necessarily distinct) elements of S, then their product, ab, is not an element of S. What is the maximum number of elements of S?

13. Some of the squares of a 3×3 board are colored black in such a way that in each row and in each column there is at least one black square. Regarding symmetries of the same coloring as distinct (i.e., considering arrangements like the two shown below as different), find the total number of ways in which one can color the board.

14. In a round robin tournament, occasionally a cyclic 3-set occurs, that is, a set $\{a, b, c\}$ of teams where a beats b, b beats c, and c beats a. If 23 teams play a round robin (without ties), what is the greatest number of cyclic 3-sets that can occur?

15. Define a *normal series* to be an expression $a_1 \pm a_2 \pm a_3 \pm \ldots \pm a_n$, where each $\pm$ is either a $+$ or a $-$, and the operations are performed in the normal order from left to right. Define a *reverse series* to be an expression $R : a_1 \pm a_2 \pm a_3 \pm \ldots \pm a_n$, where again each $\pm$ is either a $+$ or a $-$, but now the operations are grouped from right to left. For instance,

$$R : \ 3 - 4 - 6 + 5 = 3 - (4 - (6 + 5)) = 3 - 4 + 6 + 5 = 10.$$

As just illustrated, each reverse series is equivalent to a normal series with the same terms in the same order, but with some of the signs changed. Conversely, each normal series has an equivalent reverse series. If the normal series $a_1 \pm a_2 \pm a_3 \pm \ldots \pm a_{1000}$ has 600 minuses among its 999 signs, at most how many minuses does the equivalent reverse series have?

AIME Answer Key

1983 AIME

1. 060	6. 035	11. 288
2. 015	7. 057	12. 065
3. 020	8. 061	13. 448
4. 026	9. 012	14. 130
5. 004	10. 432	15. 175

1984 AIME

1. 093	6. 024	11. 106
2. 592	7. 997	12. 401
3. 144	8. 160	13. 015
4. 649	9. 020	14. 038
5. 512	10. 119	15. 036

1985 AIME

1. 384	6. 315	11. 085
2. 026	7. 757	12. 182
3. 198	8. 061	13. 401
4. 032	9. 049	14. 025
5. 986	10. 600	15. 864

1986 AIME

1. 337	6. 033	11. 816
2. 104	7. 981	12. 061
3. 150	8. 141	13. 560
4. 181	9. 306	14. 750
5. 890	10. 358	15. 400

1987 AIME

1. 300	6. 193	11. 486
2. 137	7. 070	12. 019
3. 182	8. 112	13. 931
4. 480	9. 033	14. 373
5. 588	10. 120	15. 462

1988 AIME

1. 770	6. 142	11. 163
2. 169	7. 110	12. 441
3. 027	8. 364	13. 987
4. 020	9. 192	14. 084
5. 634	10. 840	15. 704

Dropped AIME

1. 028	6. 048	11. 993
2. 274	7. 172	12. 969
3. 734	8. 540	13. 265
4. 396	9. 170	14. 506
5. 133	10. 463	15. 799

1983 AIME Solutions

1. (60) Converting each of the given logarithms into exponential form gives $x^{24} = w$, $y^{40} = w$, $(xyz)^{12} = w$. It follows that

$$z^{12} = \frac{w}{x^{12}y^{12}} = \frac{w}{w^{1/2}w^{3/10}} = w^{1/5}.$$

Thus $w = z^{60}$ and $\log_z w = 60$.

2. (15) Since $0 < p \le x \le 15$, then $|x-p| = x-p$, $|x-15| = 15-x$, and $|x - (p+15)| = p + 15 - x$. Thus

$$f(x) = (x-p) + (15-x) + (p + 15 - x) = 30 - x.$$

It follows that $f(x)$ is least when x is greatest, and that the answer is 15.

3. (20) Substitute to simplify. Several choices work well; the following substitution, which eliminates the radical immediately, is perhaps best. Define u to be the nonnegative number such that $u^2 = x^2 + 18x + 45$. (There is such a u, for if $x^2 + 18x + 45$ were negative, the right-hand side of the original equation would be undefined.) So

$$u^2 - 15 = 2\sqrt{u^2} = 2u \quad \text{(since } u \ge 0\text{)},$$
$$u^2 - 2u - 15 = 0,$$
$$(u-5)(u+3) = 0.$$

Since $u \ge 0$, we have $u = 5$. That is, x is a solution of the original equation iff (if and only if) $x^2 + 18x + 45 = 5^2$, i.e., iff $x^2 + 18x + 20 = 0$. Both solutions to this last equation are real (why?) and their product is the constant term, 20.

Note. By being careful about the sign of u and by using iff-arguments we have avoided introducing any extraneous roots for x.

4. (26)   We introduce a coordinate system as shown in the figure. WLOG (Without Loss Of Generality) we have put the top corner of the notch in the First Quadrant and have positioned the sides of the notch parallel to the axes. Letting $A = (u, v)$ it follows, using the given dimensions, that $B = (u, v-6)$ and $C = (u+2, v-6)$. Our problem is to determine $u^2 + (v-6)^2$. Since A and C are

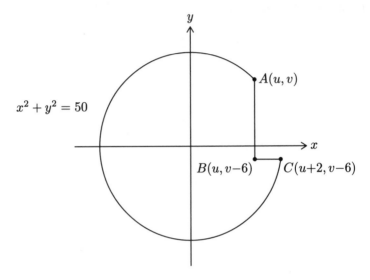

both on the circle $x^2 + y^2 = 50$, we may substitute their coordinates into this equation to get the system

$$u^2 + v^2 = 50,$$
$$(u + 2)^2 + (v-6)^2 = 50.$$

Subtracting the first equation from the second leads to $u - 3v + 10 = 0$. Solving for u and substituting into the first equation gives

$$(3v - 10)^2 + v^2 = 50,$$
$$v^2 - 6v + 5 = 0,$$
$$v = 1, u = -7 \quad \text{or} \quad v = 5, \ u = 5.$$

Since $A = (u, v)$ is in the First Quadrant, we have $A = (5, 5)$ and $B = (5, -1)$. Finally, the square of the distance from B to the origin is $25 + 1 = 26$.

5. (4) We are given that $x^2 + y^2 = 7$ and $x^3 + y^3 = 10$. Because we are asked to find the sum $x + y$, rather than x or y individually, we are moved to rewrite these equations so as to exhibit the sum:

$$x^2 + y^2 = (x+y)^2 - 2xy = 7,$$
$$x^3 + y^3 = (x+y)^3 - 3xy(x+y) = 10. \tag{1}$$

This further suggests defining

$$s = x + y, \quad p = xy. \tag{2}$$

Substituting (2) into (1) gives

$$s^2 - 2p = 7,$$
$$s^3 - 3ps = 10. \tag{3}$$

Solving for p in the first line of (3) and substituting the result into the second line gives

$$s^3 - 3s\left(\frac{s^2 - 7}{2}\right) = 10 \iff s^3 - 21s + 20 = 0. \tag{4}$$

An obvious root of this is $s = 1$. Dividing the cubic by $s - 1$ gives the factorization $(s-1)(s^2+s-20) = 0$ and thence

$$(s - 1)(s - 4)(s + 5) = 0.$$

Thus *all* values of $s = x + y$ are real and the largest is 4.

There is a gap in this argument. Clearly any solution to (1) has led by substitution to some solution to (4), but we must also show that the largest solution for s in (4) is one which arises this way. We show more, that every solution for s in (4) arises this way. This converse is hardly automatic, as the equations are not linear.

Given any s satisfying (4), set $p = \dfrac{s^2 - 7}{2}$. Then (4) $\implies$ (3). Next, for this pair (s, p), there necessarily exists a pair (x, y) satisfying (2), namely, the (possibly complex) roots of $z^2 - sz + p = 0$. Finally, substituting (2) into (3) gets us back to (1). That is, there is a solution (x, y) to (1) with $x + y = s$.

Alternative Solution. We use the method of Newton sums for determining

$$S_n = x^n + y^n, \quad n \geq 0,$$

where x and y are any two complex numbers. Setting $s = x + y$, $p = xy$, we have again that x and y are the roots of $z^2 - sz + p = 0$. We claim that

$$S_{n+2} - sS_{n+1} + pS_n = 0, \quad n \geq 0. \tag{1}$$

This is shown by adding $x^n(x^2 - sx + p) = x^n \cdot 0 = 0$ and $y^n(y^2 - sy + p) = 0$. Now let x and y be the desired solutions to

$$x^2 + y^2 = 7, \quad x^3 + y^3 = 10. \tag{2}$$

That is, we are given $S_2 = 7$ and $S_3 = 10$. We seek $S_1 = s$. We also know that $S_0 = x^0 + y^0 = 2$. Thus setting $n = 0$ and then $n = 1$ in (1), we obtain

$$\begin{aligned} 7 - s^2 + 2p &= 0, \\ 10 - 7s + ps &= 0. \end{aligned} \tag{3}$$

Thus $p = \dfrac{s^2 - 7}{2}$ and

$$10 - 7s + \frac{s^3 - 7s}{2} = 0 \iff s^3 - 21s + 20 = 0. \tag{4}$$

Thus, as before, $s = -5, 1, 4$ and the largest value is 4.

Also as before, the argument is not complete: every solution to (2) yields a solution to (4), but we also want the converse. The converse can be proved by methods similar to those in the first solution but a little more involved. Can the reader do it?

Third Solution. Since we are asked to find the largest *real* value of $s = x + y$, it suffices to find all x and y of the form $x = \frac{1}{2}(s + bi)$, $y = \frac{1}{2}(s - bi)$, where s and b are real. (If fact, as the previous solutions show, all x and y that solve $x^2 + y^2 = 7$, $x^3 + y^3 = 10$ are of this form.) Expanding powers and grouping gives

$$7 = x^2 + y^2 = \frac{1}{2}(s^2 - b^2),$$

$$10 = x^3 + y^3 = \frac{1}{4}(s^3 - 3sb^2).$$

To eliminate b, multiply the first equation above by $6s$, the second by 4, and subtract. The result is $s^3 - 21s + 20 = 0$, as before. Also as before, one can show that every solution for s comes from solutions for x and y. In fact, to obtain $s = 4$, the values for x and y are $\frac{1}{2}(4 \pm i\sqrt{2})$.

6. (35) For n odd we may write

$$
\begin{aligned}
a_n &= (7-1)^n + (7+1)^n \\
&= \left(7^n - \tbinom{n}{1}7^{n-1} + \cdots - 1\right) + \left(7^n + \tbinom{n}{1}7^{n-1} + \cdots + 1\right) \\
&= 2\left(7^n + \tbinom{n}{2}7^{n-2} + \cdots + \tbinom{n}{n-3}7^3 + \tbinom{n}{n-1}7\right) \\
&= 2 \cdot 49\left(7^{n-2} + \tbinom{n}{2}7^{n-4} + \cdots + \tbinom{n}{n-3}7\right) + 14n.
\end{aligned}
$$

It follows that $a_{83} = 49k + 14 \cdot 83 = 49k + 1162$, where k is an integer. Thus, the remainder on dividing a_{83} by 49 is the same as the remainder on dividing 1162 by 49. That remainder is 35.

7. (57) It is perhaps easier to think in terms of n knights, where $n > 4$. We observe first that there are $\binom{n}{3}$ ways of selecting 3 knights if there are no restrictions. But how many of these threesomes include at least two table neighbors?

First, there are n ways to pick three neighboring knights. (Consider each knight along with the two knights to his immediate right.) Second, there are n ways to pick two neighboring knights (as with three) followed by $(n-4)$ ways of picking a third non-neighboring knight. (We must avoid the pair and the two knights on either side.) Thus, there are $n(n-4)$ threesomes that include exactly two neighbors.

Letting P_n be the probability that at least two of the three chosen knights had been neighbors, it follows that

$$
P_n = \frac{n + n(n-4)}{\binom{n}{3}} = \frac{6(n-3)}{(n-1)(n-2)}.
$$

Then $P = P_{25} = \frac{11}{46}$ and the required sum is 57.

8. (61)   Let p be a prime less than 100. Then

$$n = \frac{1 \cdot 2 \cdot 3 \cdots p \cdots 2p \cdots 3p \cdots kp \cdots 200}{(1 \cdot 2 \cdot 3 \cdots p \cdots 2p \cdots \cdots jp \cdots 100)^2}.$$

Thus, if in addition $p^2 > 200$, we have

$$n = (\text{a } p\text{-free integer}) \cdot \frac{p^k}{p^{2j}}.$$

Now, such a prime divides n iff $k > 2j$. Once we show that there is at least one prime meeting these conditions, our answer will be the largest such prime, because any p with $p^2 \leq 200$ will be smaller. Since $kp \leq 200$, the requirement that p be as large as possible leads to our choosing k as small as possible. Thus we take $k = 3$ and $j = 1$, for which the largest p is 61. Since $61^2 > 200$, the second display above is correct for $p = 61$. Thus 61 divides n and is the largest 2-digit prime to do so.

Alternative Solution. Observe that

$$n = \binom{200}{100} = \frac{200!}{(100!)^2} = \frac{101 \cdot 102 \cdot 103 \cdots 200}{1 \cdot 2 \cdot 3 \cdots 100}$$

$$= \frac{101 \cdot 103 \cdot 105 \cdots 199}{1 \cdot 2 \cdot 3 \cdots 50} \cdot 2^{50},$$

where the denominator divides the numerator, but in doing so it cannot remove any prime greater than 50. Now, no odd integer can be twice a prime, but it might be three times a prime. Since $3 \cdot 67 > 199$, the biggest two-digit prime remaining as a prime factor of n is 61, found in the factor $183 = 3 \cdot 61$.

9. (12)   Dividing out gives

$$f(x) = 9x \sin x + \frac{4}{x \sin x}.$$

Call the first term on the right u, the second v, and note that uv is constant. This suggests using the geometric-arithmetic mean inequality:

$$\frac{u + v}{2} \geq \sqrt{uv},$$

where u, v are any nonnegative numbers and equality holds iff $u = v$. Applying this inequality to our u and v gives

$$\frac{f(x)}{2} \geq \sqrt{9 \cdot 4}, \quad \text{or} \quad f(x) \geq 12.$$

The value 12 is actually attained iff there is an x for which

$$9x \sin x = \frac{4}{x \sin x}, \quad \text{i.e.,} \quad x^2 \sin^2 x = \frac{4}{9}.$$

Since $x^2 \sin^2 x$ is 0 when $x = 0$ and it exceeds 1 when $x = \pi/2$, it follows that it equals the intermediate value $4/9$ somewhere between 0 and $\pi/2$. Thus the minimum value of $f(x)$ is indeed 12.

Remark. Unfortunately, the problem can also be solved by an almost straight-forward application of calculus. Letting $g(u) = 9u + 4/u$, where $u = x \sin x$, one finds that $g'(u) = 0$ when $u = \pm 2/3$, and that $g(2/3) = 12$. (Since $0 < x < \pi$ implies that $u > 0$, the other root can be discarded.) However, if one does not make the substitution $u = x \sin x$, but instead differentiates $f(x)$ in the form given in the problem, then a calculus approach is very painful!

10. (432)   We split the set of numbers of the type described into two subsets and make two separate subcounts:

   1. Those numbers having two 1's. The second may be placed in any one of three positions and the two other numbers (distinct) may be placed in $9 \cdot 8$ ways. Thus, there are $3 \cdot 9 \cdot 8 = 216$ such numbers.

   2. Those numbers having two of some digit other than 1. The pair of identical digits may be selected in 9 ways and then placed in 3 ways (positions 2 and 3, 2 and 4, or 3 and 4). The remaining number may be selected in 8 ways. Thus there are $9 \cdot 3 \cdot 8 = 216$ such numbers.

Thus, the answer is $216 + 216 = 432$. (It is coincidental, perhaps remarkable, that the two subtotals are the same. This is not so for, say, the 5-digit version of this problem. Try it!)

Alternative Solution. Consider 4 cells (for the 4-digit number) as below:

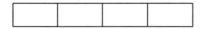

Select two of the cells as the ones that are to receive the equal digits. Call the left of these cells L and the right one R. This can be done in $\binom{4}{2} = 6$ ways. For instance, one might have

	L	R	

Starting with the first cell, fill in all cells except R with distinct numbers. This can be done in $1 \cdot 9 \cdot 8$ ways (since there is only one choice, the number 1, for the first cell, which is never R). Finally, fill R with the same number as was inserted in L. This can be done in just 1 way. Thus, the "product of ways" is $6 \cdot 1 \cdot 9 \cdot 8 \cdot 1 = 432$.

11. (288) Consider the regular tetrahedron $ABCD$, shown in the figure, that has edge length $2s$. Connect the midpoints of AC, BC, AD and BD. The quadrilateral thus determined is a square (why?) and the plane of this square divides the tetrahedron into two solids that are identical with the solid given in the problem. Thus, we have only to find the volume of the tetrahedron and then divide by 2.

The formula for the volume of a regular tetrahedron of side length e (its derivation requires only right-triangle trigonometry) is

$$V = \frac{\sqrt{2}}{12} e^3.$$

Thus, substituting $e = 2s = 12\sqrt{2}$ and taking $\frac{1}{2}V$ gives 288.

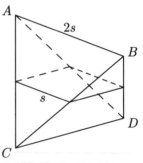

12. (65) Let $AB = 10t + u$, where t and u are digits. Then $CD = 10u + t$. We must now find t and u for which OH will be a positive rational. Evidently, $t \geq u$; to insure that OH is positive we take $t > u$.

Since $\triangle OCH$ is a right triangle, we may use the Pythagorean Theorem to express OH in terms of AB and CD:

$$OH = \sqrt{(OC)^2 - (CH)^2}$$
$$= \frac{1}{2}\sqrt{(10t + u)^2 - (10u + t)^2}$$
$$= \frac{3}{2}\sqrt{11(t^2 - u^2)}.$$

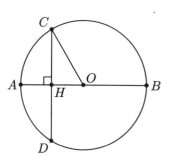

It follows that OH is rational iff $\sqrt{11(t^2-u^2)}$ is rational. But the square root of an integer is rational only if it is integral. Thus, we must find t and u for which $11(t^2-u^2)$ is a perfect square, and this will be the case only if there is a positive integer m such that

$$t^2 - u^2 = 11m^2,$$
$$(t - u)(t + u) = 11m^2.$$

Now 11 cannot divide $t - u$. (Why?) Therefore, it must divide $t + u$. But this is possible only if $t + u$ is 11. (Why?) It follows that $t - u = m^2$. Thus, we seek two numbers whose sum is 11 and whose difference is a perfect square.

Finally, we examine all t and u $(t > u)$ for which $t + u = 11$: $t = 9, u = 2; t = 8, u = 3; t = 7, u = 4;$ and $t = 6, u = 5$. Only in the last case is $t - u$ a perfect square. Thus $AB = 65$.

Note. Without the condition that OH is positive rational, the problem has many answers. For example, take $AB = 52$ and $CD = 25$ (OH is irrational), or $AB = 77 = CD$ (OH is zero). It happens that (in base ten) there is just one case in which OH is a positive rational.

13. (448) It is easier, perhaps, to generalize the problem (ever so slightly) by considering the alternating sums for *all* subsets of $\{1, 2, 3, \ldots, n\}$ that is, by including the empty set. To include the empty set without affecting the answer we have only to declare that its alternating sum be 0. The subsets of $\{1, 2, 3, \ldots, n\}$, may be divided into two kinds: those that do not contain n and those that do. Moreover, each subset of the first kind may be paired – in a one-to-one correspondence – with a subset of the second kind as follows:

$$\{a_1, a_2, a_3, \ldots, a_i\} \longleftrightarrow \{n, a_1, a_2, a_3, \ldots, a_i\}.$$

(For the empty set we have the correspondence $\phi \longleftrightarrow \{n\}$.) Then, assuming $n > a_1 > a_2 > \cdots > a_i$, the sum of the alternating sums for each such pair of subsets is given by

$$(a_1 - a_2 + a_3 - \cdots \pm a_i) + (n - a_1 + a_2 - a_3 + \cdots \mp a_i) = n.$$

And since there are 2^{n-1} such pairs of subsets (why?), the required sum is $n2^{n-1}$. Finally, taking $n = 7$, we obtain 448.

Remark. One can also conjecture the formula $f(n) = n2^{n-1}$ by computing $f(1) = 1$, $f(2) = 4$, $f(3) = 12$, and $f(4) = 32$, and then observing that $1 = 1 \cdot 1$, $4 = 2 \cdot 2$, $12 = 3 \cdot 4$, and $32 = 4 \cdot 8$. Clearly, there are many other patterns which can be fitted to the numbers $1, 4, 12, 32$, hence this "method of solution" is just a "wild guess", which happens to work. Nevertheless, it can prompt one to think of pairing the subsets of $\{1, 2, 3, \ldots, n\}$, and producing n from each pair of subsets.

14. (130) We complete the figure as shown. Applying the Pythagorean Theorem to the shaded triangle gives

$$(2a)^2 + (b - c)^2 = 12^2.$$

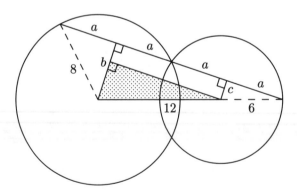

But $b = \sqrt{64 - a^2}$ and $c = \sqrt{36 - a^2}$. Thus

$$4a^2 + \left(\sqrt{64 - a^2} - \sqrt{36 - a^2}\right)^2 = 144.$$

Simplifying gives $a^2 - 22 = \sqrt{(64 - a^2)(36 - a^2)}$. Squaring and simplifying gives $4a^2 = 130$. Thus $(2a)^2 = 4a^2 = 130$.

Alternative Solution. Let X and Y be the centers of the given circles of radii x and y, respectively, assume that X and Y are z units apart, and let P be one of the two points at which the two circles intersect one another. Reflect one of the circles, say the one whose center is Y, in P (this has the same effect as a $180°$ rotation at P), and let Z be the center of this third circle, as shown in the following figure.

It is given that $QP = PR$ and that Q is on the circle centered at X. By the central symmetry at P, Q is also on the circle centered at Z. We will find $(PQ)^2$ by first determining $u = XZ$, from $\triangle XYP$ and $\triangle XPZ$, and then we will find $a = SP$, in right triangle $\triangle PSZ$, where S is the point at which PQ and XZ intersect. For simplicity, we will also let $\theta = \angle XPY$ (and hence $\angle XPZ = \pi - \theta$) and $\alpha = \angle XZP$.

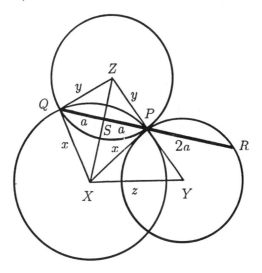

By the Law of Cosines in $\triangle XYP$ and $\triangle XPZ$,

$$z^2 = x^2 + y^2 - 2xy\cos\theta$$

and

$$u^2 = x^2 + y^2 - 2xy\cos(\pi - \theta) = x^2 + y^2 + 2xy\cos\theta.$$

Upon adding these equations, we find that

$$u^2 = 2x^2 + 2y^2 - z^2.$$

Moreover, in $\triangle XPZ$,

$$\cos \alpha = (y^2 + u^2 - x^2)/2uy.$$

With the given data, these equations yield $u = \sqrt{56}$ and $\cos \alpha = \sqrt{7}/(6\sqrt{2})$. Therefore, $\sin \alpha = \sqrt{1 - \cos^2 \alpha} = \sqrt{65/72}$, and from right triangle $\triangle PSZ$, $a = y \sin \alpha$, and hence $(PQ)^2 = 4a^2 = 4y^2 \sin^2 \alpha = 130$.

Remark. One of the advantages of the Alternative Solution is that it provides a method for constructing the desired chords of the circles, and hence it proves their existence and uniqueness for any two intersecting circles. A second advantage of it is that it sheds some light on how the problem was devised: We started with three circles, and then removed the third one and posed a question about the other two. Thirdly, with the help of the formulas derived and a bit more work, one can show that in general, $PQ = 4\Delta/u$, where Δ is the area of $\triangle XYP$ and $u = XZ$.

Unfortunately, the data for the problem was chosen rather poorly, since one can show that for the given x, y, z, the point R is on the extension of XY. Therefore, it was possible to "derive" the right answer by the following "solution", called to our attention by Gwen Kyle, a teacher at Bearden High School in Knoxville, Tennessee:

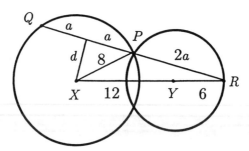

On the basis of the figure above, we find that $d^2 + a^2 = 8^2$ and $d^2 + (a + 2a)^2 = (12 + 6)^2$. From these it follows that $8^2 - a^2 = 18^2 - 9a^2$, and hence $(2a)^2 = \frac{1}{2}(18^2 - 8^2) = 130$.

15. (175) Consider the family of all chords emanating from A. Then the locus of the endpoints of these chords is the given circle of radius 5 and the locus of their midpoints is an internally tangent circle (at A) of radius $5/2$. (We have simply shrunk the given circle relative to A by a factor of $1/2$.) Now BC must be tangent to the smaller circle. For if it cut that circle twice, as in Figure 1, then there would be two chords emanating from A that are bisected by BC.

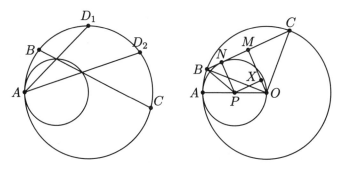

Figure 1   Figure 2

Next consider Figure 2, where O and P are the centers of the two circles and BC is tangent to the smaller circle (at N) as described before. We introduce the following construction lines: OA (through P), OB, OC, $MO \perp BC$, $NP \perp BC$, and $PX \parallel BC$. We seek $\sin \angle BOA$. We will find this by first finding all three sides of $\triangle BOP$. We have

1. $MC = 3$ and $MO = \sqrt{5^2 - 3^2} = 4$.
2. $NP = \frac{1}{2}AO = \frac{5}{2}$.
3. $MNPX$ is a rectangle, so $MX = \frac{5}{2}$ and $OX = \frac{3}{2}$.
4. $\triangle PXO$ is a right triangle, with $OP = \frac{5}{2}$ and $OX = \frac{3}{2}$, so $PX = \frac{4}{2} = 2$.
5. $MN = 2$ so $BN = 1$.
6. $\triangle BNP$ is a right triangle, so $BP^2 = 1 + \frac{25}{4} = \frac{29}{4}$.

We now have all three sides of $\triangle BOP$, so by the Law of Cosines applied to $\theta = \angle BOP$ we obtain

$$\frac{29}{4} = 25 + \frac{25}{4} - 2 \cdot 5 \cdot \frac{5}{2} \cos \theta,$$

$$\cos \theta = \frac{24}{25}.$$

Thus $\sin \theta = \sqrt{1 - \left(\frac{24}{25}\right)^2} = \frac{7}{25}$ and the desired answer is 175.

1984 AIME Solutions

1. (93)   The sum of an arithmetic progression is the product of the number of terms and the arithmetic average of the first and last terms. Therefore, $a_1 + a_2 + \cdots + a_{98} = 98(a_1 + a_{98})/2 = 49(a_1 + a_{98}) = 137$. Similarly, $a_2 + a_4 + \cdots + a_{98} = 49(a_2 + a_{98})/2 = 49(a_1 + 1 + a_{98})/2 = (49(a_1 + a_{98})/2) + (49/2) = (137/2) + (49/2) = 93$.

   Alternative Solution. Separating the terms of odd and even subscripts, let $S_o = a_1 + a_3 + \cdots + a_{97}$ and $S_e = a_2 + a_4 + \cdots + a_{98}$. Then each of these series has 49 terms, and since $a_{n+1} = a_n + 1$ for all $n \geq 1$, we have $S_o = S_e - 49$. Furthermore, $S_e + S_o = 137$. Substituting for S_o in the last equation, and solving for S_e, one finds that $S_e = 93$.

2. (592)   Note that n is a common multiple of 5 and 3. As a multiple of 5, n must end in 0, as the digit 5 is not allowed. As a multiple of 3, n must contain a number of 8's equal to a multiple of 3. Hence, in view of the minimality requirement, $n = 8880$, and $8880/15 = 592$ is the answer to the problem.

3. (144)   Let T denote the area of $\triangle ABC$, and denote by T_1, T_2 and T_3 the areas of t_1, t_2 and t_3, respectively. Moreover, let c be the length of AB, and let c_1, c_2 and c_3 be the lengths of the bases parallel to AB of t_1, t_2, and t_3, respectively. Then, in view of the similarity of the four triangles, one has

$$\frac{\sqrt{T_1}}{\sqrt{T}} = \frac{c_1}{c}, \quad \frac{\sqrt{T_2}}{\sqrt{T}} = \frac{c_2}{c} \quad \text{and} \quad \frac{\sqrt{T_3}}{\sqrt{T}} = \frac{c_3}{c}.$$

Moreover, since $c_1 + c_2 + c_3 = c$, it follows that

192

$$\frac{\sqrt{T_1}}{\sqrt{T}} + \frac{\sqrt{T_2}}{\sqrt{T}} + \frac{\sqrt{T_3}}{\sqrt{T}} = \frac{c_1 + c_2 + c_3}{c} = 1,$$

and therefore, $T = (\sqrt{T_1} + \sqrt{T_2} + \sqrt{T_3})^2 = (\sqrt{4} + \sqrt{9} + \sqrt{49})^2 = 144$.

Remark. Interestingly, this problem was also posed — in a more general setting — in Hungary's mathematics competition for specialized mathematical high schools at about the same time. The contestants there found that, even if P is not interior to $\triangle ABC$, one can find the area of $\triangle ABC$ from t_1, t_2, and t_3. More specifically, see the figures below and the corresponding formulas, where T denotes the area of $\triangle ABC$. By the way, our AIME problem was later posed as an exercise in *Középiskolai Matematikai Lapok (KöMaL* — Hungary's high school mathematics journal), in Issue 6/1985.

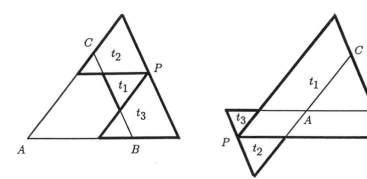

$$\sqrt{T} = -\sqrt{t_1} + \sqrt{t_2} + \sqrt{t_3} \qquad \sqrt{T} = \sqrt{t_1} - \sqrt{t_2} - \sqrt{t_3}$$

It turns out that there is also a three–dimensional extension of these results, and one can show that if P is an interior point of a tetrahedron of volume V, and if the volumes of the four tetrahedra cut off by planes parallel to the faces of the tetrahedron through P are v_1, v_2, v_3, and v_4, then

$$\sqrt[3]{V} = \sqrt[3]{v_1} + \sqrt[3]{v_2} + \sqrt[3]{v_3} + \sqrt[3]{v_4}.$$

This problem was posed as *Gy* (Exercise) 2813 in Issue 1992/10 of *KöMaL*. Can the reader adjust the above relation to cover the cases when P is not interior to the tetrahedron? What about higher dimensions?

4. (649)   Let n denote the number of positive integers in S, and let m denote their sum. Then, on the basis of the information given,

$$\frac{m}{n} = 56 \quad \text{and} \quad \frac{m - 68}{n - 1} = 55.$$

Solving these equations simultaneously yields $n = 13$ and $m = 728$. Now, to maximize the largest element of S, one must minimize the others. To attain this, S must contain eleven 1's, a 68 and a 649, since $728 - 11 - 68 = 649$.

5. (512)   Adding the two equations and using standard log properties yields

$$\log_8 a + \log_8 b + \log_4 a^2 + \log_4 b^2 = \log_8(ab) + 2\log_4(ab) = 12.$$

Moreover, since $\log_8 x = (\log_2 x)/(\log_2 8) = \frac{1}{3}\log_2 x$, and similarly, $\log_4 x = (\log_2 x)/(\log_2 4) = \frac{1}{2}\log_2 x$, the above equation is equivalent to

$$\frac{4}{3}\log_2(ab) = 12.$$

It follows that $\log_2(ab) = 9$, and hence $ab = 2^9 = 512$.

6. (24)   First observe that the three circles are disjoint, i.e. their centers are more than 6 units apart. Next note that any line through (17,76) will divide the area of the circle centered there evenly. Thus the problem reduces to finding the value of m for which the line given by

$$y - 76 = m(x-17) \tag{1}$$

also divides the areas of the other two circles in the desired manner. To this end, note that such a line must pass exactly as far to the right of (14,92) as it passes to the left of (19,84). Denoting this distance by h, it follows that $(14+h, 92)$ and $(19-h, 84)$ must satisfy Equation (1), i.e.,

$$\frac{16}{h-3} = m = \frac{8}{2-h}, \qquad (2)$$

from which $h = \frac{7}{3}$, $m = -24$ and the answer to the problem is 24.

Note. As shown in the first figure below, in this case the uniqueness of the solution is assured by the fact that $h < 3$, the common length of the circles' radii. If the circles were positioned differently with respect to one another, or if the value of h were larger than 3, as exemplified in the second figure below, then the resulting line would not necessarily yield a unique solution to the problem.

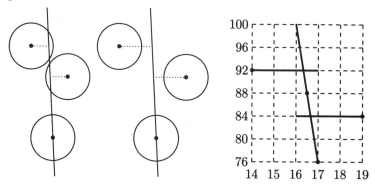

Alternative Solution. Note that the lines passing through (16.5,88), the center of symmetry of the two circles centered at (14,92) and (19,84), divide the areas of these two circles in the manner desired. Consequently, the line through (16.5,88) and (17,76), whose slope is $(88 - 76)/(16.5 - 17) = -24$, provides the answer to the problem. The uniqueness of the solution follows from observing that it intersects the radius from (14,92) to (17,92) of one circle and the radius from (16,84) to (19,84) of the other circle, as shown in the third figure above. (Note the difference in scales.)

7. (997) Rather than assaulting $f(84)$ directly, it is advantageous to start with values of $f(n)$ near $n = 1000$, and search for a pattern when n is less than 1000, thereby utilizing the recursive definition of f.

Indeed, one finds that

$$f(999) = f(f(1004)) = f(1001) = 998,$$
$$f(998) = f(f(1003)) = f(1000) = 997,$$
$$f(997) = f(f(1002)) = f(999) = 998, \qquad (1)$$
$$f(996) = f(f(1001)) = f(998) = 997,$$
$$f(995) = f(f(1000)) = f(997) = 998,$$

on the basis of which one may conjecture that

$$f(n) = \begin{cases} 997, & \text{if } n \text{ is even and } n < 1000, \\ 998, & \text{if } n \text{ is odd and } n < 1000. \end{cases} \qquad (2)$$

To prove (2), it is convenient to use downward induction. In this, the inductive step is to prove (2) for n, assuming that it is true for all m, $n < m < 1000$. Since the definition relates $f(n)$ to $f(n+5)$, we can do the inductive step only when $n + 5 < 1000$, that is, we must verify the truth of (2) for $n = 999, 998, \ldots, 995$ separately. This was done in (1). Now for $n < 995$,

$$f(n) = f(f(n+5)) = \begin{cases} f(997) = 998, & \text{if } n + 5 \text{ is even,} \\ f(998) = 997, & \text{if } n + 5 \text{ is odd.} \end{cases}$$

Noting that n is even when $n + 5$ is odd, and that n is odd when $n + 5$ is even, completes the proof. In particular, it follows from (2) that $f(84) = 997$.

8. (160)   Let $w = z^3$. Then the given equation reduces to

$$w^2 + w + 1 = 0,$$

whose solutions are $(-1 + i\sqrt{3})/2$ and $(-1 - i\sqrt{3})/2$, with arguments of $120°$ and $240°$, respectively. From these, one finds the following six values for the argument of z:

$$\frac{120°}{3}, \quad \frac{120° + 360°}{3}, \quad \frac{120° + 720°}{3},$$

$$\frac{240°}{3}, \quad \frac{240° + 360°}{3}, \quad \frac{240° + 720°}{3}.$$

Clearly, only the second one of these, $160°$, is between $90°$ and $180°$.

Alternative Solution. Multiplying the given equation by $z^3 - 1 = 0$ yields

$$z^9 - 1 = 0,$$

whose solutions are the ninth roots of unity:

$$z_n = \cos(n \cdot 40°) + i \sin(n \cdot 40°), \quad n = 0, 1, 2, \ldots, 8.$$

Of these, only z_3 and z_4 are in the second quadrant. However, since the solutions of $z^9 - 1 = 0$ are distinct, and since z_3 is a solution of $z^3 - 1 = 0$, it cannot be a solution of the original equation. It follows that the desired root is z_4, with degree measure 160.

9. (20) Let V be the volume of tetrahedron $ABCD$ and h the altitude of the tetrahedron corresponding to D. Then $V = h \cdot \text{area}(\triangle ABC)/3$, which will be determined upon finding h.

By definition, for the angle between faces ABC and ABD to be 30°, planes perpendicular to AB must cut the two faces in rays which form a 30° angle. Choose such a plane through D, let K be its intersection with AB, and let H be the point on the line of intersection of the plane chosen and the plane 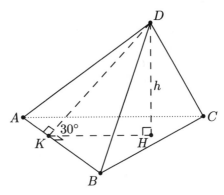 of $\triangle ABC$ so that $DH \perp KH$. These are shown in the adjoining figure. Since $DK \perp AB$, we find that

$$DK = 2 \cdot \text{area}(\triangle ABD)/AB = 8 \text{ cm}.$$

Moreover, since $\triangle DKH$ is a 30° − 60° − 90° triangle, it follows that $h = DH = DK/2 = 4$cm. Consequently, substituting into the formula given in the first paragraph, we find that $V = 20$cm^3.

10. (119)   Given that $s = 30 + 4c - w > 80$,   the problem calls for finding the lowest value of s for which the corresponding value of c is unique. To this end, first observe that if $c + w \leq 25$, then by increasing the number of correct answers by 1 and the number of wrong answers by 4, one attains the same score. (This is made possible by the equivalence of the above inequality to $(c + 1) + (w + 4) \leq 30$). Consequently, one must have

$$c + w \geq 26. \tag{1}$$

Next observe that

$$w \leq 3, \tag{2}$$

for otherwise one could reduce the number of wrong answers by 4 and the number of correct answers by 1, and still attain the same score. (The latter is possible since $s > 80$ clearly implies that $c \geq 13$.) Now to minimize s, we must minimize c and maximize w, subject to Inequalities (1) and (2) above. This leads to $w = 3$, $c = 23$ and $s = 30 + 4 \cdot 23 - 3 = 119$.

11. (106)   The 12 trees can be planted in 12! orders. Let k be the number of orders in which no two birch trees are adjacent to one another. The probability we need is $k/(12!)$. To find k, we will count the number of patterns

$$\underline{\quad} N \underline{\quad} N \underline{\quad} N \underline{\quad} N \underline{\quad} N \underline{\quad} N \underline{\quad} N \underline{\quad}$$
$$\quad 1 \qquad 2 \qquad 3 \qquad 4 \qquad 5 \qquad 6 \qquad 7 \qquad 8$$

where the 7 N's denote nonbirch (i.e., maple and oak) trees, and slots 1 through 8 are to be occupied by birch trees, at most one in each slot. There are 7! orders for the nonbirch trees, and for each ordering of them there are $8 \cdot 7 \cdot 6 \cdot 5 \cdot 4$ ways to place the birch trees. Thus, we find that $k = (7!) \cdot 8 \cdot 7 \cdot 6 \cdot 5 \cdot 4$,

$$\frac{m}{n} = \frac{(7!) \cdot 8 \cdot 7 \cdot 6 \cdot 5 \cdot 4}{12!} = \frac{7}{99} \text{ and } m + n = 106.$$

Note. We have assumed in this solution that each tree is distinguishable. The problem can also be interpreted to mean that trees are distinguishable if and only if they are of different species. In that case, the calculations (i.e., the numerator and the denominator) are different, but the probability turns out to be the same.

12. (401) First we use the given equations to find various numbers in the domain to which f assigns the same values. We find that

$$f(x) = f(2 + (x-2)) = f(2 - (x-2)) = f(4-x) \qquad (1)$$

and that

$$f(4-x) = f(7 - (x+3)) = f(7 + (x+3)) = f(x+10). \qquad (2)$$

From Equations (1) and (2) it follows that

$$f(x+10) = f(x). \qquad (3)$$

Replacing x by $x + 10$ and then by $x - 10$ in Equation (3), we get $f(x+10) = f(x+20)$ and $f(x-10) = f(x)$. Continuing in this way, it follows that

$$f(x+10n) = f(x), \quad \text{for} \quad n = \pm 1, \pm 2, \pm 3, \ldots. \qquad (4)$$

Since $f(0) = 0$. Equation (4) implies that

$$f(\pm 10) = f(\pm 20) = \cdots = f(\pm 1000) = 0,$$

necessitating a total of 201 roots for the equation $f(x) = 0$ in the closed interval $[-1000, \ 1000]$.

Next note that by setting $x = 0$ in Equation (1), $f(4) = f(0) = 0$ follows. Therefore, setting $x = 4$ in Equation (4), we obtain 200 more roots for $f(x) = 0$ at $x = -996, -986, \ldots, -6, 4, 14, \ldots, 994$. Since the zig-zag function pictured below satisfies the given conditons and has precisely these and no other roots, the answer to the problem is 401.

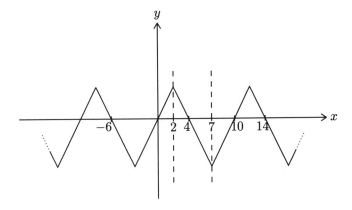

Note 1. Geometrically, the conditions imply symmetry with respect to the lines $x = 2$ and $x = 7$. The solution above shows how to draw consequences from these symmetries through algebra. More generally, one can prove that if a function is symmetric with respect to $x = a$ and $x = b > a$, then it must also have translational symmetry for every integral multiple of $2(b - a)$. This fact can also be proven geometrically, leading to an alternate formulation of the solution.

Note 2. See the related AHSME problem #16 from 1984.

13. (15) In order to simplify the notation, let

$$a = \cot^{-1} 3, \quad b = \cot^{-1} 7, \quad c = \cot^{-1} 13, \quad d = \cot^{-1} 21, \quad (1)$$

$$e = a + b \quad \text{and} \quad f = c + d. \quad (2)$$

Then, in view of (1) and (2) above, the problem is equivalent to finding the value of $10 \cot(e+f)$.

It is easy to show that in general,

$$\cot(x+y) = \frac{(\cot x)(\cot y) - 1}{\cot x + \cot y}. \quad (3)$$

Moreover, even if $\cot^{-1}$ is viewed as a multiple valued function,

$$\cot(\cot^{-1} x) = x, \quad \text{for all real } x. \quad (4)$$

Utilizing Equations (1) through (4) above, we find that

$$\cot e = \cot(a+b) = \frac{3 \cdot 7 - 1}{3 + 7} = 2,$$

$$\cot f = \cot(c+d) = \frac{13 \cdot 21 - 1}{13 + 21} = 8,$$

$$\cot(e+f) = \frac{2 \cdot 8 - 1}{2 + 8} = \frac{3}{2},$$

and therefore, $10 \cot(e+f) = 15$ is the answer to the problem.

Note. More generally, since the numbers 3, 7, 13 and 21 are all of the form $1 + n + n^2$, one may attempt to express $\cot^{-1}(1+n+n^2)$ more advantageously. Indeed, it is not difficult to show that

$$\cot^{-1}(1+n+n^2) = \tan^{-1}(n+1) - \tan^{-1} n$$

within an integral multiple of π. Consequently, if k is a positive integer, then

$$\sum_{n=1}^{k} \cot^{-1}(1+n+n^2) = \tan^{-1}(k+1) - \tan^{-1} 1,$$

within an integral multiple of π. From this, one can prove that

$$\cot\left(\sum_{n=1}^{k} \cot^{-1}(1+n+n^2)\right) = \frac{k+2}{k}.$$

14. (38)   We will show that if k is an even integer and if $k \geq 40$, then k is expressible as the sum of two composites. This leaves 38 as the candidate for the largest even integer not expressible in such manner; it is easy to check that 38 indeed satisfies this requirement. The proof of our claim for $k \geq 40$ hinges on the fact that if n is odd and greater than 1, then $5n$ is an odd composite ending in 5. So, to express k as desired, it suffices to find small odd composites ending in 5, 7, 9, 1 and 3, and to add these to numbers of the form $5n$. Indeed, 15, 27, 9, 21 and 33 will satisfy the above condition, and in each of the following cases one can find an odd integer n, $n > 1$, such that

if k ends in 0   (i.e. $40, 50, \ldots$), then $k = 15 + 5n$,

if k ends in 2   (i.e. $42, 52, \ldots$), then $k = 27 + 5n$,

if k ends in 4   (i.e. $44, 54, \ldots$), then $k = 9 + 5n$,

if k ends in 6   (i.e. $46, 56, \ldots$), then $k = 21 + 5n$,

if k ends in 8   (i.e. $48, 58, \ldots$), then $k = 33 + 5n$.

Alternative Solution.   First observe that $6n + 9$ is an odd composite number for $n = 0, 1, 2, \ldots$. Now partition the set of even positive integers into three residue classes, modulo 6, and note that in each of the following cases the indicated decompositions are satisfied by some nonnegative integer, n:

if $k \equiv 0 \pmod 6$ and $k \geq 18$, then $k = 9 + (6n + 9)$,

if $k \equiv 2 \pmod 6$ and $k \geq 44$, then $k = 35 + (6n + 9)$,

if $k \equiv 4 \pmod 6$ and $k \geq 34$, then $k = 25 + (6n + 9)$.

This takes care of all even integers greater than 38. Checking again shows that 38 is the answer to the problem.

15. (36)   The claim that the given system of equations is satisfied by x^2, y^2, z^2 and w^2 is equivalent to claiming that

$$\frac{x^2}{t-1} + \frac{y^2}{t-9} + \frac{z^2}{t-25} + \frac{w^2}{t-49} = 1 \qquad (1)$$

is satisfied by $t = 4$, 16, 36 and 64. Multiplying to clear fractions, we find that for all values of t for which it is defined (i.e., $t \neq 1$, 9, 25 and 49), Equation (1) is equivalent to the polynomial equation

$$\begin{aligned}
(t-1)(t-9)(t-25)(t-49) & \\
- x^2(t-9)(t-25)(t-49) &- y^2(t-1)(t-25)(t-49) \qquad (2) \\
- z^2(t-1)(t-9)(t-49) &- w^2(t-1)(t-9)(t-25) = 0,
\end{aligned}$$

where the left member may be viewed as a fourth degree polynomial in t. Since $t = 4$, 16, 36 and 64 are known to be roots, and a fourth degree polynomial can have at most 4 roots, these must be all the roots. It follows that (2) is equivalent to

$$(t-4)(t-16)(t-36)(t-64) = 0. \qquad (3)$$

Since the coefficient of t^4 is 1 in both (2) and (3), one may conclude that the coefficients of the other powers of t must also be the same. In particular, equating the coefficients of t^3, we have

$$1 + 9 + 25 + 49 + x^2 + y^2 + z^2 + w^2 = 4 + 16 + 36 + 64,$$

from which $x^2 + y^2 + z^2 + w^2 = 36$.

Note. By equating the left members of Equations (2) and (3) and letting $t = 1$, one also finds that $x^2 = 11025/1024$. Similarly, letting $t = 9$, 25 and 49 in succession yields $y^2 = 10395/1024$, $z^2 = 9009/1024$ and $w^2 = 6435/1024$. One can show that these values indeed satisfy the given system of equations, and that their sum is 36.

Remark. By observing that

$$\frac{x^2}{2^2 - 1^2} = 1 \qquad \text{implies} \qquad x^2 = 3$$

and that

$$\frac{x^2}{2^2 - 1^2} + \frac{y^2}{2^2 - 3^2} = 1 \quad \text{and} \quad \frac{x^2}{4^2 - 1^2} + \frac{y^2}{4^2 - 3^2} = 1$$

$$\text{imply} \quad x^2 + y^2 = 10,$$

and by writing 3 as 1·3 and 10 as 2·5, one might guess that the next term of the sequence is 3·7 = 21, and that the answer to the present problem is 4·9 = 36. This shortcut was called to our attention by Andrew Chin, a former USAMO winner, who obtained his Ph.D. from the University of Oxford as a Rhodes Scholar and an NSF Fellow. Clearly, the evidence for this pattern is at best shaky, but in mathematics one must usually conjecture prior to proving. It so happens that the pattern is valid, and can be established by the method given in the solution above.

1985 AIME Solutions

1. (384) For each $n > 1$, $x_{n-1}x_n = n$. Using this fact for $n = 2, 4$, 6 and 8, one obtains $x_1x_2 \cdots x_8 = (x_1x_2)(x_3x_4)(x_5x_6)(x_7x_8) = 2 \cdot 4 \cdot 6 \cdot 8 = 384$.

   Note. Except for the fact that it must be nonzero, the value of x_1 does not affect the solution.

2. (26) The volume of a cone with a circular base of radius r and height h is given by $\frac{\pi}{3}hr^2$. Denoting the length of the legs of the right triangle by a and b, this implies that

   $$\frac{\pi}{3}ba^2 = 800\pi \qquad \text{and} \qquad \frac{\pi}{3}ab^2 = 1920\pi.$$

   Dividing the first of these equations by the second one yields $\frac{a}{b} = \frac{5}{12}$. Hence, $a = \frac{5}{12}b$ and $a^3 = \frac{5}{12}ba^2 = \frac{5}{12}(800)(3) = 1000$. It follows that $a = 10, b = 24$ and that the triangle's hypotenuse (by Pythagoras' Theorem) is 26 cm.

3. (198) First note that one may write c in the form

   $$c = a(a^2 - 3b^2) + i[b(3a^2 - b^2) - 107]. \tag{1}$$

   From this, in view of the fact that c is real, one may conclude that

   $$b(3a^2 - b^2) = 107. \tag{2}$$

   Since a and b are positive integers, and since 107 is prime, two possible cases arise from (2):

   $$\text{either} \qquad b = 107 \quad \text{and} \quad 3a^2 - b^2 = 1,$$
   $$\text{or} \qquad b = 1 \quad \text{and} \quad 3a^2 - b^2 = 107.$$

   In the first case, $3a^2 = 107^2 + 1$ would follow. But this is impossible, since $107^2 + 1$ is not a multiple of 3.
   In the second case, one finds that $a = 6$ and hence, in view of (1), $c = a(a^2 - 3b^2) = 6(6^2 - 3 \cdot 1^2) = 198$.

4. (32) Let $A, B, \ldots, H$ be the points labeled in the adjoining figure, with FH parallel to EB. Noting that $\triangle FGH$ is similar to $\triangle DCE$, it then follows that

204

$$\frac{FG}{FH} = \frac{DC}{DE},$$

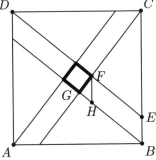

from which

$$FG^2 = \frac{DC^2}{DE^2} \cdot FH^2.$$

In view of the given information, this is equivalent to

$$\frac{1}{1985} = \frac{1}{1 + (1 - \frac{1}{n})^2} \cdot \frac{1}{n^2},$$

from which

$$2n^2 - 2n + 1 = 1985.$$

Since this last equation is equivalent to $2(n - 32)(n + 31) = 0$, and since n is positive, it follows that $n = 32$.

Remark. Based on the present problem, a somewhat more difficult problem was posed in the Fall 1988 issue of the *Pi Mu Epsilon Journal.* It asked for the characterization of those positive integers n for which the side of the inscribed square is the reciprocal of an integer. It can be shown that n must be of the form

$$\frac{1}{4}((1 + \sqrt{2})^k + (1 - \sqrt{2})^k + 2)$$

where k is an odd positive integer.

5. (986) Calculating the first eight terms of the sequence, one finds that it cycles in blocks of six terms; i.e., for $n = 1, 2, 3, \ldots, a_{n+6} = a_n$. More specifically,

$$a_n = \begin{cases} a_1 & \text{if } n = 1, 7, 13, \ldots, \\ a_2 & \text{if } n = 2, 8, 14, \ldots, \\ a_2 - a_1 & \text{if } n = 3, 9, 15, \ldots, \\ -a_1 & \text{if } n = 4, 10, 16, \ldots, \\ -a_2 & \text{if } n = 5, 11, 17, \ldots, \\ a_1 - a_2 & \text{if } n = 6, 12, 18, \ldots . \end{cases}$$

Since the sum of any six consecutive terms of the sequence is zero, if we let s_n be the sum of the first n terms, then

$$s_n = \begin{cases} a_1 & \text{if } n = 1, 7, 13, \ldots, \\ a_1 + a_2 & \text{if } n = 2, 8, 14, \ldots, \\ 2a_2 & \text{if } n = 3, 9, 15, \ldots, 2001, \ldots, \\ 2a_2 - a_1 & \text{if } n = 4, 10, 16, \ldots, 1492, \ldots, \\ a_2 - a_1 & \text{if } n = 5, 11, 17, \ldots, 1985, \ldots, \\ 0 & \text{if } n = 6, 12, 18, \ldots. \end{cases}$$

Therefore,

$$s_{1985} = a_2 - a_1 = 1492$$

and

$$s_{1492} = 2a_2 - a_1 = 1985,$$

from which $a_2 = 493$ and $s_{2001} = 2a_2 = 986$.

6. (315)   The key to the solution is the fact that if two triangles have the same height, then their areas are proportional to their bases. Therefore, from the figure below, we have the equations

$$\frac{40}{30} = \frac{40 + y + 84}{30 + 35 + x}, \quad \frac{35}{x} = \frac{35 + 30 + 40}{x + 84 + y} \quad \text{and} \quad \frac{84}{y} = \frac{84 + x + 35}{y + 40 + 30}.$$

Solving the first two of these equations simultaneously, one finds that $x = 70$ and $y = 56$. After checking that these values also satisfy the third equation, one may conclude that the area of $\triangle ABC$ is $30 + 35 + 70 + 84 + 56 + 40$ or 315.

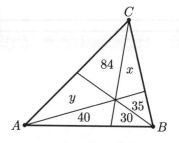

Remark. Since x and y had to satisfy all three equations featured in the solution, it took a bit of effort to choose the data given in the problem. Therefore, it might be instructive to sketch our process for the setting of this problem.

To this end, assume that the Cevians through P (that is, the three lines from a vertex through P to the opposite side) divide the sides of $\triangle ABC$ into segments of lengths a, b, c, d, e, and f and that the distances of P from sides AB, BC, and CA are $2u$, $2v$, and $2w$, respectively. Then the areas

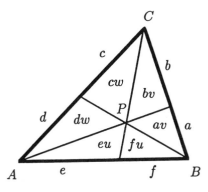

of the smaller triangles surrounding P are as shown in the figure on the right. They must satisfy the three equations

$$\frac{cw + dw}{eu} = \frac{bv + av}{fu}, \quad \frac{eu + fu}{av} = \frac{dw + cw}{bv}, \quad \frac{av + bv}{cw} = \frac{fu + eu}{dw},$$

which can be simplified to

$$f(c+d)w = e(b+a)v, \quad b(e+f)u = a(d+c)w, \quad d(a+b)v = c(f+e)u.$$

Observe now that in view of Ceva's Theorem,

$$\frac{a}{b} \cdot \frac{c}{d} \cdot \frac{e}{f} = 1,$$

the third of these equations is a consequence of the first two; i.e., $d(a+b)v = \frac{ac}{bf}e(b+a)v = \frac{ac}{bf}f(c+d)w = \frac{c}{b}a(d+c)w = \frac{c}{b}b(e+f)u = c(f+e)u$.

At this point we observed that if $d = a$, $e = b$, and $f = c$, then Ceva's Theorem is satisfied as long as $a + b$, $c + a$, $b + c$ satisfy the triangle inequality, which is the case if a, b, c satisfy it. Hence we chose $a = 2$, $b = 4$, $c = 3$, reducing the first two equations to $15w = 24v$ and $28u = 10w$, one of whose solutions led to the data featured in the problem.

An additional benefit of this development was that by applying the "scaling factor" $t = \sqrt{5}\sqrt{7}\sqrt[4]{2}/2$ (found via Heron's Formula) to the sides of the 5–6–7 triangle, we could actually exhibit a triangle with a point P in it, having the desired properties.

This problem, along with the next, was featured in the October 1986 issue of the South African *Mathematical Digest*, where they credited *Elementa*, a Swedish magazine for both problems. Hopefully, *Elementa* credited the AIME!

7. (757)   Since the prime factorization of positive integers is unique, and since 4 is relatively prime to 5 and 2 is relatively prime to 3, one may conclude that there exist positive integers m and n such that

$$a = m^4, \quad b = m^5, \quad c = n^2 \quad \text{and} \quad d = n^3.$$

Then

$$19 = c - a = n^2 - m^4 = (n - m^2)(n + m^2).$$

Since 19 is a prime, and since $n - m^2 < n + m^2$, it follows that

$$n - m^2 = 1 \quad \text{and} \quad n + m^2 = 19.$$

Therefore, $m = 3$, $n = 10$, $d = 1000$, $b = 243$ and $d - b = 1000 - 243 = 757$.

8. (61)   Two preliminary observations are needed:

(i) Each A_i should be 2 or 3, because for any scheme meeting this condition, $M < 1$; while for any other choice of the A_i's, $M > 1$.

(ii) There is only one way to sum seven integers, each of them 2 or 3, and to obtain 19: two of them must be 2, while the other five must be 3.

In view of the above, and to make M as small as possible, one must round down (to 2) the two numbers with smallest decimal parts (i.e., a_1 and a_2), and round up (to 3) the other five a_i's. Thereby one finds that

$$M = \max\{|a_2 - 2|, |a_3 - 3|\} = \max\{.61, .35\} = .61,$$

and that $100M = 61$.

Note. More generally, one can show that if any a_i's are listed so that their decimal parts are in increasing order, if the sum of the a_i's is D, and if the sum of their integral parts is I, then the last $D - I$ of the a_i's should be rounded up (to the next integer), while the others should be rounded down.

The reader may wish to apply this procedure to the set $\{1.42, 2.39, 2.60, 3.37, 3.97, 4.45, 5.33, 5.47\}$, which were the data in an alternative formulation of this problem. We decided in favor of the data that appeared on the Exam since they allowed for the second observation given in the solution, and hence for an alternative (and easier, but much more limited) method of solution.

In any event, such unusual methods of rounding are necessary, for example, in the apportionment of congressional seats to the individual states in the United States. (The U.S. Constitution requires each state to be represented in proportion to its fraction of the U.S. population, but each state must also have an integer number of representatives, and the total size of the House is fixed.) For informative discussions and other solutions to the apportionment problem the reader is advised to read the following article in *The American Mathematical Monthly:* M. L. Balinski and H. P. Young, The Quota Method of Apportionment, Vol. 82 (1975), pp. 701-730.

9. (49)   Since any two chords of equal length subtend equal angles, the parallelism of the chords is irrelevant. Therefore, we may choose points A, B and C, as shown in the first accompanying figure, so that $AB = 2$, $BC = 3$ and (since $\widehat{AC} = \alpha + \beta$) $AC = 4$. Since $\angle ACB = \alpha/2$, by the Law of Cosines we have

$$\cos \frac{\alpha}{2} = \frac{AC^2 + BC^2 - AB^2}{2 \cdot AC \cdot BC} = \frac{4^2 + 3^2 - 2^2}{2 \cdot 4 \cdot 3} = \frac{7}{8}.$$

Therefore, one finds that

$$\cos \alpha = 2\cos^2 \frac{\alpha}{2} - 1 = 2 \cdot \frac{49}{64} - 1 = \frac{17}{32},$$

and that the desired answer is $17 + 32$ or 49.

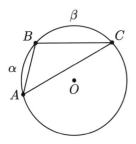
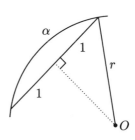

Alternative Solution. Attacking the problem more directly, let r denote the radius of the circle, and note that $\sin \frac{\alpha}{2} = \frac{1}{r}$, as shown in the second figure above. One similarly finds that $\sin \frac{\beta}{2} = \frac{3}{2r}$ and $\sin \frac{\alpha+\beta}{2} = \frac{2}{r}$. Since $\sin(\frac{\alpha}{2} + \frac{\beta}{2}) = \sin \frac{\alpha}{2} \cos \frac{\beta}{2} + \cos \frac{\alpha}{2} \sin \frac{\beta}{2}$, it follows that

$$\frac{2}{r} = \frac{1}{r}\sqrt{1 - \frac{9}{4r^2}} + \frac{3}{2r}\sqrt{1 - \frac{1}{r^2}}. \tag{1}$$

Solving this for $\frac{1}{r^2}$, one finds that it is equal to $\frac{15}{64}$. Upon checking that this root is not extraneous to (1), it follows that

$$\cos\alpha = 1 - 2\sin^2\frac{\alpha}{2} = 1 - 2\cdot\frac{15}{64} = \frac{17}{32},$$

as in the first solution.

10. (600)   Introduce the notation

$$f(x) = \lfloor 2x \rfloor + \lfloor 4x \rfloor + \lfloor 6x \rfloor + \lfloor 8x \rfloor, \tag{1}$$

and observe that if n is a positive integer, then from (1)

$$f(x + n) = f(x) + 20n \tag{2}$$

follows. In particular, this means that if an integer k can be expressed in the form $f(x_0)$ for some real number x_0, then for $n = 1, 2, 3, \ldots$ one can express $k + 20n$ similarly; i.e., $k + 20n = f(x_0) + 20n = f(x_0 + n)$. In view of this, one may restrict attention to determining which of the first 20 positive integers are generated by $f(x)$ as x ranges through the half-open interval $(0, 1]$.

Next observe that as x increases, the value of $f(x)$ changes only when either $2x, 4x, 6x$ or $8x$ attains an integral value, and that the change in $f(x)$ is always to a new, higher value. In the interval $(0,1]$ such changes occur precisely when x is of the form m/n, where $1 \le m \le n$ and $n = 2, 4, 6,$ or 8. There are 12 such fractions; in increasing order they are:

$$\frac{1}{8}, \frac{1}{6}, \frac{1}{4}, \frac{1}{3}, \frac{3}{8}, \frac{1}{2}, \frac{5}{8}, \frac{2}{3}, \frac{3}{4}, \frac{5}{6}, \frac{7}{8} \text{ and } 1.$$

Therefore, only 12 of the first 20 positive integers can be represented in the desired form. Since $1000 = (50)(20)$, in view of (2), this implies that in each of the 50 sequences,

$$1, 2, 3, \ldots, 20; \ 21, 22, 23, \ldots, 40; \ \ldots; 981, 982, 983, \ldots, 1000,$$

of 20 consecutive integers only 12 can be so expressed, leading to a total of $(50)(12)$ or 600 positive integers of the desired form.

11. (85) Let F_1 and F_2 denote the given foci, $(9,20)$ and $(49,55)$, respectively, and let $\hat{F}_2 : (49, -55)$ be the reflection of F_2 in the x-axis. We claim that k, the length of the major axis of the ellipse, is equal to the length of the segment $F_1 \hat{F}_2$, which is easily found to be 85 by the distance formula between two points in the plane.

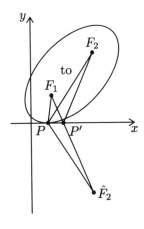

To prove our claim, let P be the point of tangency of the ellipse to the x-axis, and assume that the segment $F_1 \hat{F}_2$ intersects the x-axis at the point P', distinct from P, as shown in the adjacent figure, where the locations of the foci are purposefully distorted. Then, by the Triangle Inequality, one finds that

$$PF_1 + P\hat{F}_2 > F_1\hat{F}_2 = P'F_1 + P'\hat{F}_2 = P'F_1 + P'F_2. \qquad (1)$$

However, by the definition of the ellipse, for every point Q on the ellipse, $QF_1 + QF_2 = k$. Hence, in particular,

$$PF_1 + PF_2 = k. \qquad (2)$$

Moreover, for every point Q outside the ellipse, one must have $QF_1 + QF_2 > k$. Since, by the definition of tangency, P' must lie outside the ellipse, in particular it follows that

$$P'F_1 + P'F_2 > k. \qquad (3)$$

From (2) and (3) one may therefore conclude that

$$PF_1 + P\hat{F}_2 = PF_1 + PF_2 < P'F_1 + P'F_2,$$

which is contrary to (1). This contradiction establishes the falsity of the assumption that P and P' are distinct, and hence completes the proof of the claim.

Remark. This problem could also be solved by using calculus, as was pointed out by Lawrence Hood, a teacher at Miami Killian Senior High School in Miami, Florida. We tried to select problems so that students with less advanced course work would not be at a disadvantage. Fortunately, that was the case here too, since the following calculus solution is by no means easier than the one presented above:

Let $(x, 0)$ be any point on the x-axis, and let

$$l = \sqrt{(x - 9)^2 + 20^2} + \sqrt{(x - 49)^2 + 55^2}$$

be the sum of the distances from the foci of the ellipse to $(x, 0)$. Then the key to the solution is to recognize that l is minimized if and only if $(x, 0)$ is the point of tangency of the ellipse to the x-axis, and that the corresponding l is the length of the major axis. Setting the derivative of l (with respect to x) equal to 0 and solving the resulting quadratic equation, we find that $x = -97/7$ or $59/3$. The first of these solutions is extraneous, while the second one yields $l = 85$, the answer to the present problem.

12. (182) For $n = 0, 1, 2, \ldots,$ let a_n be the probability that the bug is at vertex A after crawling exactly n meters. Then

$$a_{n+1} = \frac{1}{3}(1 - a_n), \qquad (1)$$

because the bug can be at vertex A after crawling $n + 1$ meters if and only if

(i) it was not at A following a crawl of n meters (this has probability $1 - a_n$), and

(ii) from one of the other vertices it heads toward A (this has probability $\frac{1}{3}$).

Now since $a_0 = 1$ (i.e., the bug starts at vertex A), from (1) we have

$$a_1 = 0, \quad a_2 = \frac{1}{3}, \quad a_3 = \frac{2}{9}, \quad a_4 = \frac{7}{27}, \quad a_5 = \frac{20}{81}, \quad a_6 = \frac{61}{243},$$

and $p = a_7 = \frac{182}{729}$, leading to 182 as the answer to the problem.

Note. The above calculations may be somewhat simplified by observing that for $n \geq 1$ the numerators of a_n and a_{n+1} sum to 3^{n-1}. Alternately, one may write (1) in the form

$$a_n - \frac{1}{4} = -\frac{1}{3}\left(a_{n-1} - \frac{1}{4}\right),$$

which (upon successive substitutions, starting with $a_0 - \frac{1}{4} = \frac{3}{4}$) yields the explicit formula

$$a_n = \frac{1}{4} + \left(-\frac{1}{3}\right)^n\left(\frac{3}{4}\right).$$

For a somewhat related problem the reader is referred to the solution of E 1752 (proposed by Arthur Engel, the former coach of the West German IMO team) on Page 200 in the February 1966 issue of *The American Mathematical Monthly*.

13. (401) More generally, we will show that if a is a positive integer and if d_n is the greatest common divisor of $a+n^2$ and $a+(n+1)^2$, then the maximum value of d_n is $4a+1$, attained when $n = 2a$. It will follow that for the sequence under consideration the answer is $4(100) + 1$ or 401.

To prove the above, note that if d_n divides $a + (n + 1)^2$ and $a + n^2$, then it also divides their difference; i.e.,

$$d_n \,|\, (2n + 1). \tag{1}$$

Now, since $2(a + n^2) = n(2n + 1) + (2a - n)$, it follows from (1) that

$$d_n \,|\, (2a - n). \tag{2}$$

Hence from (1) and (2), $d_n \,|\, ((2n + 1) + 2(2a - n))$, or

$$d_n \,|\, (4a + 1). \tag{3}$$

Consequently, $1 \le d_n \le 4a + 1$, so $4a + 1$ is the largest possible value of d_n. It is attained, since for $n = 2a$ we have

$$a + n^2 = a + (2a)^2 = a(4a + 1)$$

and

$$a + (n + 1)^2 = a + (2a + 1)^2 = (a + 1)(4a + 1).$$

14. (25) Assume that a total of n players participated in the tournament. We will obtain two expressions in n: one by considering the total number of points gathered by all of the players, and one by considering the number of points gathered by the losers (10 lowest scoring contestants) and those gathered by the winners (other $n - 10$ contestants) separately. To obtain the desired expressions, we will use the fact that if k players played against one another, then they played a total of $k(k-1)/2$ games, resulting in a total of $k(k - 1)/2$ points to be shared among them.

In view of the last observation, the n players gathered a total of $n(n-1)/2$ points in the tournament. Similarly, the losers had $10 \cdot 9/2$ or 45 points in games among themselves; since this accounts for half of their points, they must have had a total of 90 points. In games among themselves the $n-10$ winners similarly gathered $(n-10)(n-11)/2$ points; this also accounts for only half of their total number of points (the other half coming from games against the losers), so their total was $(n-10)(n-11)$ points. Thus we have the equation

$$n(n-1)/2 = 90 + (n-10)(n-11),$$

which is equivalent to

$$n^2 - 41n + 400 = 0.$$

Since the left member of this equation may be factored as $(n-16)(n-25)$, it follows that $n = 16$ or 25. We discard the first of these in view of the following observation: if there were only 16 players in the tournament, then there would have been only 6 winners, and the total of their points would have been 30 points, resulting in an average of 5 points for each of them. This is less than the $90/10$ or 9 points gathered, on the average, by each of the losers! Therefore, $n = 25$; i.e., there were 25 players in the tournament.

Remark. To ensure the legitimacy of the problem, one should be able to exhibit at least one tournament with 25 players which satisfies the given conditions. To this end, denote the players by

$$A_1, A_2, \ldots, A_6, B_1, B_2, \ldots, B_9, C_1, C_2, \ldots, C_{10}.$$

Then if

- the A's tie among themselves, win six times and tie three times against the B's (say, A_1, A_2 tie B_1, B_2, B_3, A_3, A_4 tie B_4, B_5, B_6, and A_5, A_6 tie B_7, B_8, B_9), and win against the C's;
- the B's tie among themselves and tie against the C's;
- the C's tie among themselves,

then each of the A's gathers 20 points (10 against the C's), each of the B's gathers 10 points (5 against the C's, and each of the C's gathers 9 points ($4\frac{1}{2}$ against the C's, who are clearly the 10 players with the least number of points).

Query. Can the reader construct another tournament meeting all the conditions but not involving any ties?

15. (864)   At each of the vertices P, Q, R and S, marked in the first figure below, each of the three facial angles measures $90°$. Consequently, the polyhedron may be immediately recognized as a portion of a cube. Moreover, two such polyhedra may be fitted together along their hexagonal faces. To accomplish this, flip the second polyhedron and rotate it by $60°$ (as shown in the second figure below), and then slide the two polyhedra together until X and X', Y and Y', as well as Z and Z' coincide. As shown in the third figure below, the result of the procedure is a cube of volume 12^3 or 1728 cm^3. Therefore, the volume of the polyhedron under consideration is $1728/2$ or 864 cm^3.

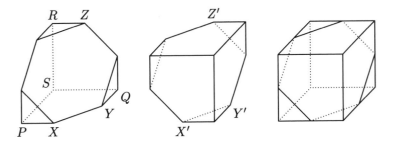

1986 AIME Solutions

1. (337) Let $y = \sqrt[4]{x}$. Then the equation may be written in the form

$$y^2 - 7y + 12 = 0,$$

whose roots are $y = 3$ and $y = 4$. Consequently, we obtain the x-values of 3^4 and 4^4, whose sum is 337.

2. (104) Repeated use of the identity $(x + y)(x - y) = x^2 - y^2$ leads to

$$(\sqrt{5} + \sqrt{6} + \sqrt{7})(\sqrt{5} + \sqrt{6} - \sqrt{7}) = (\sqrt{5} + \sqrt{6})^2 - (\sqrt{7})^2$$
$$= (11 + 2\sqrt{30}) - 7 = 4 + 2\sqrt{30},$$
$$(\sqrt{5} - \sqrt{6} + \sqrt{7})(-\sqrt{5} + \sqrt{6} + \sqrt{7}) = (\sqrt{7})^2 - (\sqrt{5} - \sqrt{6})^2$$
$$= 7 - (11 - 2\sqrt{30}) = -4 + 2\sqrt{30},$$

and

$$(4 + 2\sqrt{30})(-4 + 2\sqrt{30}) = (2\sqrt{30})^2 - 4^2$$
$$= 120 - 16 = 104.$$

Alternative Solution. In view of Heron's Formula, one may observe that the given expression is the square of the area of a triangle whose sides are $2\sqrt{5}, 2\sqrt{6}, 2\sqrt{7}$. Upon computing that the altitude of this triangle corresponding to the base $2\sqrt{7}$ is $2\sqrt{26}/\sqrt{7}$, the result follows.

3. (150) Multiplying both sides of $\cot x + \cot y = 30$ by $\tan x \tan y$, we find that

$$\tan y + \tan x = 30 \tan x \tan y.$$

Since the left side of this equation is equal to 25, it follows that

$$\tan x \tan y = \frac{25}{30} = \frac{5}{6}.$$

Therefore,

$$\tan(x + y) = \frac{\tan x + \tan y}{1 - \tan x \tan y} = \frac{25}{1 - \frac{5}{6}} = 150.$$

4. (181) Adding the five given equations, and then dividing both sides of the resulting equation by 6, yields

$$x_1 + x_2 + x_3 + x_4 + x_5 = 31. \qquad (1)$$

Subtracting (1) from the given fourth and fifth equations, we find that $x_4 = 17$ and $x_5 = 65$. Consequently, $3x_4 + 2x_5 = 51 + 130 = 181$.

5. (890) By division we find that

$$n^3 + 100 = (n + 10)(n^2 - 10n + 100) - 900.$$

Thus, if $n + 10$ divides $n^3 + 100$, then it must also divide 900. Moreover, since n is maximized whenever $n + 10$ is, and since the largest divisor of 900 is 900, we must have $n + 10 = 900$. Therefore, $n = 890$.

6. (033) Let k be the number of the page that was counted twice. Then, $0 < k < n + 1$, and $1 + 2 + \cdots + n + k$ is between $1 + 2 + \cdots + n$ and $1 + 2 + \cdots + n + (n + 1)$. In other words, $n(n + 1)/2 < 1986 < (n + 1)(n + 2)/2$; i.e.,

$$n(n + 1) < 3972 < (n + 1)(n + 2).$$

By trial and error (clearly, n is a little larger than 60) we find that $n = 62$. Thus $k = 1986 - (62)(63)/2 = 1986 - 1953 = 33$.

7. (981) If we use only the first six non-negative integral powers of 3, namely 1, 3, 9, 27, 81 and 243, then we can form only 63 terms, since

$$\binom{6}{1} + \binom{6}{2} + \cdots + \binom{6}{6} = 2^6 - 1 = 63.$$

Consequently, the next highest power of 3, namely 729, is also needed.

After the first 63 terms of the sequence the next largest ones will have 729 but not 243 as a summand. There are 32 of these, since $\binom{5}{0} + \binom{5}{1} + \cdots + \binom{5}{5} = 32$, bringing the total number of terms to 95. Since we need the 100th term, we must next include 243 and omit 81. Doing so, we find that the $96^{th}, 97^{th}, \ldots, 100^{th}$ terms are: $729 + 243, 729 + 243 + 1, 729 + 243 + 3, 729 + 243 + 3 + 1$ and $729 + 243 + 9 = 981$.

Alternative Solution. Note that a positive integer is a term of this sequence if and only if its base 3 representation consists only of 0's and 1's. Therefore, we can set up a one-to-one correspondence between the positive integers and the terms of this sequence by representing both with binary digits (0's and 1's), first in base 2 and then in base 3:

$$1 = 1_{(2)} \longleftrightarrow 1_{(3)} = 1$$
$$2 = 10_{(2)} \longleftrightarrow 10_{(3)} = 3$$
$$3 = 11_{(2)} \longleftrightarrow 11_{(3)} = 4$$
$$4 = 100_{(2)} \longleftrightarrow 100_{(3)} = 9$$
$$5 = 101_{(2)} \longleftrightarrow 101_{(3)} = 10$$
$$\vdots$$

This is a correspondence between the two sequences in the order given, that is, the k-th positive integer is made to correspond to the k-th sum (in increasing order) of distinct powers of 3. This is because, when the binary numbers are written in increasing order, they are still in increasing order when interpreted in any other base. (If you can explain why this is true when interpreted in base 10, you should be able to explain it in base 3 as well.)

Therefore, to find the 100th term of the sequence, we need only look at the 100th line of the above correspondence.

$$100 = 1100100_{(2)} \longleftrightarrow 1100100_{(3)} = 981.$$

8. (141) The number $1000000 = (2^6)(5^6)$ has $(6+1)(6+1) = 49$ distinct positive divisors. To see this, observe that they are all of the form $(2^i)(5^j)$; thus there are seven choices for i $(0, 1, 2, \ldots, 6)$ and, independently, the same seven choices for j. Apart from 1000, the other 48 divisors form 24 pairs such that the product of each pair is 1000000. Since one of these pairs consists of the improper divisors 1 and 1000000, it follows that the product of all proper divisors of 1000000 is $(1000)(1000000)^{23}$ or 10^{141}. Moreover, since the sum, S, of the logarithms is equal to the logarithm of the product of these numbers, $S = 141$. The nearest integer to 141 is clearly 141.

Note. See the related AHSME problem #25 from 1986.

9. (306)   As shown in the figure on the right, $EH = BC - (BE + HC) = BC - (FP + PG) = 450 - d$. In like manner, $GD = 510 - d$. Moreover, from the similarity of $\triangle DPG$ and $\triangle ABC$ we have $DP/GD = AB/CA$. Hence

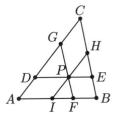

$$DP = \frac{AB}{CA} \cdot GD = \frac{425}{510}(510 - d) = 425 - \frac{5}{6}d. \qquad (1)$$

In like manner, since $\triangle PEH$ and $\triangle ABC$ are similar, $PE/EH = AB/BC$. Hence

$$PE = \frac{AB}{BC} \cdot EH = \frac{425}{450}(450 - d) = 425 - \frac{17}{18}d. \qquad (2)$$

Since $d = DP + PE$, adding (1) and (2) we find that $d = 850 - \frac{16}{9}d$, from which $d = 306$.

Alternative Solution. More generally, let $AB = c$, $BC = a$, $CA = b$, $EH = x$, $GD = y$, and $IF = z$. Then, with $DE = FG = HI = d$, we find that $DP/y = c/b$, $PE/x = c/a$, and hence $d = DE = DP + PE = c\left(\frac{y}{b} + \frac{x}{a}\right)$. From this, since $d = c - z$, it follows that $c - z = c\left(\frac{y}{b} + \frac{x}{a}\right)$; i.e., $\frac{x}{a} + \frac{y}{b} + \frac{z}{c} = 1$. Substituting $a - d$, $b - d$, and $c - d$ for x, y, and z, respectively, and rearranging the resulting equation leads to

$$\frac{1}{a} + \frac{1}{b} + \frac{1}{c} = \frac{2}{d}, \qquad (3)$$

which may be viewed as the key to setting and/or solving the problem.

To use Equation (3) for the generation of "nice" AIME–like problems, start off with carefully chosen positive integers p, q, r, and their sum, s, and divide through the equation $p + q + r = s$ by the least common multiple of p, q, r, s. To generate the data for the present problem, we started with $15 + 17 + 18 = 50$, divided through by 7650, and found that

$$\frac{1}{510} + \frac{1}{450} + \frac{1}{425} = \frac{2}{306}.$$

Since the triple $(510, 450, 425)$ satisfies the triangle inequality, this choice of data led to a healthy problem. By contrast, $(p, q, r) = (3, 8, 9)$ would not have done so.

Third Solution (sketch). A solution is more straightforward using knowledge of *barycentric coordinates.* Every point P in a triangle ABC can be represented by a triple of nonnegative numbers α, β, γ, where α is the ratio of the altitudes of $\triangle PBC$ and $\triangle ABC$, with corresponding definitions for β and γ. Equivalently, α is the ratio of the areas of $\triangle PBC$ and $\triangle ABC$. It follows that $\alpha + \beta + \gamma = 1$. Now, turning to the problem at hand, and using similar triangles and the fact that the lines through P all have length d, we find that

$$1 - \alpha = \frac{d}{BC}, \quad 1 - \beta = \frac{d}{AC}, \quad 1 - \gamma = \frac{d}{AB}.$$

Adding these equations, substituting $\alpha + \beta + \gamma = 1$, and dividing by d, we obtain

$$\frac{2}{d} = \frac{1}{BC} + \frac{1}{AC} + \frac{1}{AB},$$

the same Equation (3) we used to solve for d before.

Barycentric coordinates are often used in advanced mathematics (e.g., topology and the study of convex sets in many dimensions) where they are usually defined using linear algebra. For more about their use in classical geometry, see Dan Pedoe, "Geometry, a Comprehensive Course," Dover Publications.

10. (358)   Adding (abc) to N, and observing that each of the digits a, b, and c appears exactly twice in each column, we are led to the equation

$$N + (abc) = 222(a + b + c). \tag{1}$$

In view of this, the problem can be resolved by searching for an integral multiple of 222, say $222k$, which is larger than N (since $(abc) \neq 0$) and less that $N + 1000$ (since (abc) is a three-digit number), so that (1) is satisfied. That is, the sum of the digits of $222k - N$ must be k. If this holds, then $222k - N$ is the answer to the problem.

In our case, since $(14)(222) < 3194$ and $(19)(222) > 3194 + 1000$, the search is limited to $k = 15, 16, 17$ and 18. Of these only 16 works. Specifically, $(16)(222) - 3194 = 358$ and $3+5+8 = 16$.

Note. By also observing that $2N = 9(27a + 47b + 49c) + (a + b + c)$, one can use the following reasoning to slightly simplify the computations. With $k = a+b+c$ as before, k meets two conditions: $2N \equiv k \pmod 9$, and $N < 222k < N + 1000$. Moreover, at most five integers k satisfy the second condition, since dividing it by 222 gives

$$\frac{N}{222} < k < \frac{N}{222} + \frac{1000}{222} < \frac{N}{222} + 5.$$

We conclude that there is at most one integer k meeting both conditions. On the other hand, if the participants in the parlor game really did pick a number (abc) and did compute N accurately, then there will be at least one integer k, namely, $a+b+c$. Thus, we conclude that there is exactly one k that meets the two conditions, and the magician can find it by computing the *smallest* k that satisfies $2N \equiv k \pmod 9$ and $222k > N$. Then $(abc) = 222k - N$. Note that this argument works for any N, not just 3194 as in the problem.

This problem was based on "A Christmas party piece" by M.T.L. Bizley, which appeared as Note 59.16 in the December 1975 issue of *The Mathematical Gazette*.

11. (816) Substituting $y - 1$ for x, the given expression becomes

$$1 - (y - 1) + (y - 1)^2 - (y - 1)^3 + \cdots + (y - 1)^{16} - (y - 1)^{17},$$

which may be written in the form

$$1 + (1 - y) + (1 - y)^2 + (1 - y)^3 + \cdots + (1 - y)^{16} + (1 - y)^{17}. \quad (1)$$

Note that each $(1 - y)^k$ term in (1) will yield a y^2 term for $2 \le k \le 17$. More specifically, by the Binomial Theorem, each of the summands in (1) contributes $\binom{k}{2}$ or $k(k-1)/2$ to the coefficient of y^2. Therefore, the problem is equivalent to computing the sum

$$\binom{2}{2} + \binom{3}{2} + \binom{4}{2} + \cdots + \binom{17}{2}.$$

To this end, one may proceed directly (i.e., by calculating and adding the sixteen numbers 1, 3, 6, ..., 136); or use the result of the derivation

$$\sum_{k=2}^{n} \frac{k(k-1)}{2} = \sum_{k=1}^{n} \frac{k(k-1)}{2} = \frac{1}{2}(\sum_{k=1}^{n} k^2 - \sum_{k=1}^{n} k)$$

$$= \frac{1}{2}(\frac{n(n+1)(2n+1)}{6} - \frac{n(n+1)}{2})$$

$$= \frac{(n+1)n(n-1)}{6};$$

or use the more general formula

$$\sum_{k=m}^{n} \binom{k}{m} = \binom{n+1}{m+1}.$$

In any case, the desired sum is equal to 816.

Remark. The following alternative derivation was suggested by Dennis White, who came across the problem and devised his solution while substitute teaching at Northwest Catholic High School in West Hartford, Connecticut:

Since

$$1 - x + x^2 - x^3 + \cdots + x^{17} = \frac{1 - x^{18}}{1 + x},$$

it follows that

$$a_0 + a_1 y + a_2 y^2 + a_3 y^3 + \cdots + a_{16} y^{16} + a_{17} y^{17} = \frac{1 - (y-1)^{18}}{y}.$$

Therefore, $1 - (y-1)^{18} = a_0 y + a_1 y^2 + a_2 y^3 + \cdots + a_{16} y^{17} + a_{17} y^{18}$. Since

$$(y-1)^{18} = \sum_{k=0}^{18} \binom{18}{k}(-y)^k,$$

it follows immediately that a_2, the coefficient of y^3, is $\binom{18}{3} = 816$.

12. (061) First we show that S contains at most 5 elements. Suppose otherwise. Then S has at least $\binom{6}{1} + \binom{6}{2} + \binom{6}{3} + \binom{6}{4}$ or 56 subsets of 4 or fewer members. The sum of each of these subsets is at most 54 (since $15 + 14 + 13 + 12 = 54$), hence, by the Pigeonhole Principle, at least two of these sums are equal. If the subsets are disjoint, we are done; if not, then the removal of the common element(s) yields the desired contradiction.

Next we attempt to construct such a 5-element set S, by choosing its elements as large as possible. Including 15, 14 and 13 in S leads to no contradiction, but if 12 is also in S, then (in view of $12 + 15 = 13 + 14$) the conditions on S would be violated. Hence we must omit 12. No contradiction results from letting 11 be a member of S, but then $10 \notin S$ since $10 + 15 = 11 + 14$, and $9 \notin S$ since $9 + 15 = 11 + 13$. So we must settle for 8 as the fifth element of S. Indeed, $S = \{8, 11, 13, 14, 15\}$ satisfies the conditions of the problem, yielding $8 + 11 + 13 + 14 + 15$ or 61 as the candidate for its solution.

Finally, to show that the maximum is indeed 61, suppose that the sum is more for another choice of S. Observe that this set must also contain 15, 14 and 13, for if even the smallest of them (13) is omitted, the maximum possible sum (62) is achievable only by including 10, 11 and 12, but then $15 + 11 = 14 + 12$. Having chosen 15, 14 and 13, we must exclude 12, as noted before. If 11 is included, then we are limited to the sum of 61 as above. If 11 is not included, then even by including 10 and 9 (which we can't) we could not surpass 61 since $15 + 14 + 13 + 10 + 9 = 61$. Consequently, 61 is indeed the maximum sum one can attain.

Remark. In the original formulation of the problem, the largest element of S was restricted to be no greater than 25. The problem was posed with this data in the USA Mathematical Talent Search in the Summer 1990 issue of *Consortium.* The interested reader may also wish to consult George Berzsenyi's article, entitled "A Pigeonhole for Every Pigeon", on page 40 in the September–October 1990 issue of *Quantum* for further comments on this problem.

13. (560) Think of such sequences of coin tosses as progressions of blocks of T's and H's, to be denoted by (T) and (H), respectively. Next note that each HT and TH subsequence signifies a transition from (H) to (T) and from (T) to (H), respectively. Since there should be three of the first kind and four of the second kind in each of the sequences of 15 coin tosses, one may conclude that each such sequence is of the form

$$(T)(H)(T)(H)(T)(H)(T)(H). \tag{1}$$

Our next concern is the placement of T's and H's in their respective blocks, so as to assure that each sequence will have two HH and five TT subsequences. To this end, we will assume that each block in (1) initially contains only one member. Then, to satisfy the conditions of the problem, it will suffice to place two more H's into the (H)'s and five more T's into the (T)'s. Thus, to solve the problem, we must count the number of ways this can be accomplished.

Recall that the number of ways to put p indistinguishable balls (the extra H's and T's in our case) into q distinguishable boxes (the (H)'s and (T)'s, distinguished by their order in the sequence) is given by the formula $\binom{p+q-1}{p}$. (Students who are not familiar with this fact should verify it.) In our case, it implies that the two H's can be placed in the four (H)'s in $\binom{2+4-1}{2}$ or 10 ways, and the five T's can be placed in the four (T)'s in $\binom{5+4-1}{5}$ or 56 ways. The desired answer is the product, 560, of these numbers.

Note. Such problems as this (only harder) often occur in the study of DNA molecules.

14. (750) In order to find the volume of P, we will determine its dimensions, x, y and z. To this end, consider the rectangular parallelepiped shown in the first figure below, with vertices and side lengths as indicated. For definiteness, we labeled the box so that

$$d(BH, CD) = 2\sqrt{5},$$
$$d(BH, AE) = \frac{30}{\sqrt{13}} \tag{1}$$
$$\text{and } d(BH, AD) = \frac{15}{\sqrt{10}}.$$

where, in general, $d(RS, TU)$ denotes the distance between lines RS and TU.

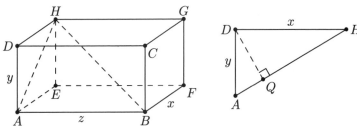

Now the distance from BH to CD is equal to the distance from CD to the plane ABH, which is the same as the length of the perpendicular from D to the diagonal AH of rectangle $AEHD$. To see this, note that this perpendicular is also perpendicular to the plane ABH and the line CD. If one "slides" it over so that its top moves from D towards C, it eventually intersects line BH. This gives an equally long segment, which is perpendicular to both CD and BH, so its length is indeed the distance $d(BH, CD)$. This distance is found via similar triangles, as shown in the second figure above, in which $DQ/DA = DH/AH$, and hence $DQ = xy/\sqrt{x^2 + y^2}$. Treating the other distances similarly, in view of (1), this leads to the equations

$$\frac{xy}{\sqrt{x^2 + y^2}} = 2\sqrt{5}, \quad \frac{yz}{\sqrt{y^2 + z^2}} = \frac{30}{\sqrt{13}} \quad \text{and} \quad \frac{zx}{\sqrt{z^2 + x^2}} = \frac{15}{\sqrt{10}}. \quad (2)$$

Upon squaring each equation in (2), taking reciprocals and simplifying one arrives at the system

$$\frac{1}{x^2} + \frac{1}{y^2} = \frac{1}{20}, \quad \frac{1}{y^2} + \frac{1}{z^2} = \frac{13}{900} \quad \text{and} \quad \frac{1}{z^2} + \frac{1}{x^2} = \frac{2}{45}. \quad (3)$$

We solve (3) by first adding the equations therein, and then subtracting from $\frac{1}{2}$ times the resulting equation each of the original equations in (3). Thus we find that $1/x^2 = 1/25$, $1/y^2 = 1/100$ and $1/z^2 = 1/225$. From these, $x = 5$, $y = 10$, and $z = 15$. Consequently, the volume of P is $xyz = 750$.

15. (400) More generally, we will show that if $\triangle ABC$ is a right triangle with right angle at C, if $AB = 2r$ and if the acute angle between the medians emanating from A and B is θ, then

$$\text{area}(\triangle ABC) = \frac{4}{3} r^2 \tan \theta. \quad (1)$$

In our case, $r = 60/2 = 30$ and $\tan \theta = 1/3$ (determined either by the standard formula for the tangent of the angle between two given lines, or by glancing at the first accompanying figure), so the answer to the problem is $(4/3)(30)^2(1/3)$ or 400.

To establish (1), first note that

$$\text{area}(\triangle ABC) = r^2 \sin \psi, \qquad (2)$$

where $\psi = \angle AOC$, as shown in the second figure below, where O is the midpoint of AB, D is the centroid of $\triangle ABC$ and $CO = r$ (since $\triangle ABC$ is a right triangle). Consequently, to prove (1), it suffices to verify that $\sin \psi = (4/3) \tan \theta$, which is equivalent to establishing that

$$\cos \phi = -(4/3) \tan \theta, \qquad (3)$$

where $\phi = \psi + 90°$. This consideration leads us to the third figure below, where O' is the center of the circle through A, D and B, and $\angle DOO' = \phi$.

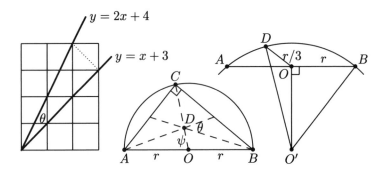

To prove (3), we will apply the Law of Cosines to $\triangle DOO'$. (The fact that $\angle OO'B = \theta$ comes from the observation that $\angle ADB = 180° - \theta$, and hence arc ADB has central angle 2θ; $DO = r/3$ is a well-known fact concerning the centroid.) Noting that $DO' = BO' = r \csc \theta$ and $OO' = r \cot \theta$ (from $\triangle BOO'$), indeed,

$$\cos \phi = \frac{(r/3)^2 + (r \cot \theta)^2 - (r \csc \theta)^2}{2(r/3)(r \cot \theta)} = -\frac{4}{3} \tan \theta,$$

as was to be shown.

Remark. Based on a suggestion of Noam Elkies (then a student at Stuyvesant High School in New York — later the youngest ever Full Professor at Harvard University), one can show more directly that if the hypotenuse of right $\triangle ABC$ is c, then its area is $(c^2 \tan \theta)/3$, where θ is the angle enclosed by its medians emanating from A and B, as shown in the figure on the right.

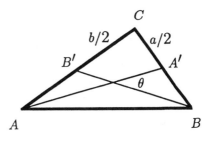

To this end, note that $\theta + \pi/2 = \angle AA'C + \angle BB'C$, and hence

$$\tan \theta = -\cot(\theta + \pi/2) = -\cot(\angle AA'C + \angle BB'C)$$
$$= \frac{(\tan \angle AA'C)(\tan \angle BB'C) - 1}{\tan \angle AA'C + \tan \angle BB'C}.$$

Since $\tan \angle AA'C = 2b/a$, $\tan \angle BB'C = 2a/b$, and $a^2 + b^2 = c^2$, the last expression above can be simplified to $3ab/2c^2$. From this, in view of the fact that area$(\triangle ABC) = ab/2$, the desired result follows.

Alternative Solution. As in the diagram shown on the right, let M, N and O be the midpoints of the sides of $\triangle ABC$, D be its centroid, p, q, s and t be the distances indicated, and $\theta = \angle ADM$. As in the previous solution, also observe that $\tan \theta = 1/3$, and hence $\sin \theta = 1/\sqrt{10}$. Then from area$(\triangle ABC) = 6 \cdot$ area$(\triangle ADM)$,

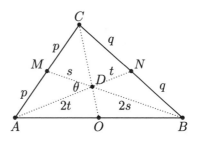

$$\text{area}(\triangle ABC) = 6st \sin \theta = \frac{6}{\sqrt{10}} st, \qquad (1)$$

and, since $\triangle ABC$ is a right triangle,

$$\text{area}(\triangle ABC) = 2pq. \qquad (2)$$

Moreover, by the Pythagorean Theorem, we find that

$$4p^2 + 4q^2 = AB^2 = 3600,$$
$$p^2 + 4q^2 = 9s^2,$$
$$4p^2 + q^2 = 9t^2.$$

With the help of these expressions, from (1) and (2) it follows that

$$\begin{aligned}
(\text{area}(\triangle ABC))^2 &= \frac{18}{5}s^2 t^2 \\
&= \frac{18}{5} \cdot \frac{p^2 + 4q^2}{9} \cdot \frac{4p^2 + q^2}{9} \\
&= \frac{2}{45}((2p^2 + 2q^2)^2 + 9p^2 q^2) \\
&= \frac{2}{45}((3600/2)^2 + \frac{9}{4}(\text{area}(\triangle ABC))^2) \\
&= 144,000 + \frac{1}{10}(\text{area}(\triangle ABC))^2.
\end{aligned}$$

Consequently, $\frac{9}{10}(\text{area}(\triangle ABC))^2 = 144,000$, from which area$(\triangle ABC) = 400$.

Third Solution. Since $y = x + 3$ and $y = 2x + 4$ intersect at $(-1, 2)$, in the diagram on the right the coordinates of A and B will be of the form $(-1+a, 2+2a)$ and $(-1+b, 2+b)$, respectively. Moreover, since -1 and 2 are the averages of the coordinates A, B, and C, it is not difficult to show that the coordinates of C are given by $(-1-a-b, 2-2a-b)$. We will find the area of $\triangle ABC$ by computing $\frac{1}{2}AC \cdot BC$ in terms of a and b, without actually determining them.

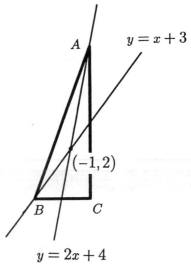

Since AC and BC are perpendicular, their slopes, $(4a+b)/(2a+b)$ and $(2a+2b)/(a+2b)$, are negative reciprocals of one another. This fact leads to the equation

$$10a^2 + 15ab + 4b^2 = 0.$$

Now note that the vector (or change) from A to C is $(2a+b, 4a+b)$, and that from B to C is $(a+2b, 2a+2b)$. Since $\triangle ABC$ is a right triangle with $AB = 60$, by the Pythagorean Theorem we also conclude that

$$(AC)^2 + (BC)^2 = \left[(2a+b)^2 + (4a+b)^2\right] + \left[(a+2b)^2 + (2a+2b)^2\right]$$
$$= 25a^2 + 24ab + 10b^2 = 3600.$$

From these two equations one can determine that $ab = -800/3$, and then use this fact to ease our computations as follows:

Since $(10a^2 + 15ab + 4b^2)^2 = 100a^4 + 300a^3b + 305a^2b^2 + 120ab^3 + 16b^4 = 0$, we find that

$$(\text{area}(\triangle ABC))^2 = \frac{1}{4}(AC)^2(BC)^2$$
$$= \frac{1}{4}[(2a+b)^2 + (4a+b)^2][(a+2b)^2 + (2a+2b)^2]$$
$$= \frac{1}{4}(100a^4 + 300a^3b + 314a^2b^2 + 120ab^3 + 16b^4)$$
$$= \frac{1}{4} \cdot 9a^2b^2 = \frac{9}{4}(-800/3)^2 = 400^2.$$

Hence the answer to our problem is 400.

Fourth solution (sktech). This approach uses an idea from the paired AHSME problem, #26 of 1986, to wit, if a right triangle is "level" (has legs parallel to the x and y axes) then the slope of the medians are $m/2$ and $2m$, where m is the slope of the hyptenuse. The median slopes in the AIME problem are 1 and 2 (not in the ratio 1:4), so the triangle is not level. If we can rotate it to make it level, we can find m. Then it is not hard to show that the area is $\frac{1}{2}L^2\dfrac{|m|}{m^2+1}$, where $L = 60$ is the length of the hypotenuse.

So let θ be the angle of a rotation that makes the triangle level, and let α and β be the angles of elevation of the medians in the leveled triangle, so that $\tan\beta = 4\tan\alpha = 2m$. Then

$$\tan(\beta - \theta) = 2, \quad \tan(\alpha - \theta) = 1.$$

Using the tangent difference formula, and setting $z = \tan\theta$, one obtains two algebraic equations in m and z, from which one finds that $m = (9 \pm \sqrt{65})/4$ and thus Area = 400.

1987 AIME Solutions

1. (300) Since there is no carrying in the addition, the ones column must add to 2, the tens column to 9, the hundreds column to 4 and the thousands column to 1 for each simple ordered pair of non-negative integers summing to 1492. To get a single digit d as the sum of two digits, there are $d + 1$ ways:

$$0 + d, \ 1 + (d{-}1), \ 2 + (d{-}2), \ldots, d + 0.$$

Thus the number of simple ordered pairs of non-negative integers that sum to 1492 is $(1 + 1)(4 + 1)(9 + 1)(2 + 1) = 300$.

2. (137) Let O and $\tilde{O}$ be the centers of the two spheres, and let P and $\tilde{P}$ be the two points where the extensions of the segment $O\tilde{O}$ pierce the two spheres, respectively, so that O is between P and $\tilde{O}$, and $\tilde{O}$ is between O and $\tilde{P}$. Then the desired maximum distance is

$$P\tilde{P} = PO + O\tilde{O} + \tilde{O}\tilde{P}, \tag{1}$$

where PO and $\tilde{O}\tilde{P}$ are the given radii and $O\tilde{O}$ is found by the Distance Formula. In our case, $P\tilde{P} = 19 + 31 + 87 = 137$.

To see that (1) indeed yields the maximum distance, note that by the Triangle Inequality, for any points Q and $\tilde{Q}$ on the spheres with centers O and $\tilde{O}$, respectively,

$$Q\tilde{Q} \leq QO + O\tilde{Q} \leq QO + O\tilde{O} + \tilde{O}\tilde{Q} = PO + O\tilde{O} + \tilde{O}\tilde{P} = P\tilde{P}.$$

Note. The above solution does not depend on the position of the spheres relative to one another, as can be seen in the two configurations below, showing cross sections of the spheres by planes containing their centers.

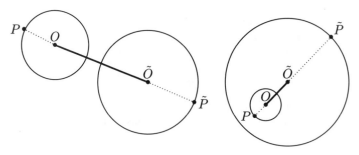

3. (182)   Let k be a positive integer, and let $1, d_1, d_2, \ldots, d_{n-1}, d_n, k$ be its divisors in ascending order. Then $1 \cdot k = d_1 \cdot d_n = d_2 \cdot d_{n-1} = \cdots$. For k to be nice, we must have $n = 2$. Moreover, d_1 must be prime, for otherwise, the proper divisors of d_1 would have appeared in the listing above between 1 and d_1. Similarly, d_2 is either a prime or the square of d_1, for otherwise d_1 could not be the only divisor between 1 and d_2. Therefore, k is either the product of two distinct primes or – being the product of a prime and its square – is the cube of a prime.

In view of the above, one can easily list the first ten nice numbers. They are: 6, 8, 10, 14, 15, 21, 22, 26, 27 and 33. Their sum is 182.

4. (480)   First we note that the graph of the given equation is symmetric with respect to the x-axis, since the replacement of y by $-y$ does not change the equation. Consequently, it suffices to assume that $y \geq 0$, find the area enclosed above the x-axis, and then double this area to find the answer to the problem.

By sketching both $y = |x/4|$ and $y = |x - 60|$ on the same graph, as shown in the first figure below, we note that $y = |x/4| - |x - 60| \geq 0$ only if the graph of $y = |x/4|$ lies above that of $y = |x - 60|$. This occurs only between $x = 48$ (where $x/4 = 60 - x$) and $x = 80$ (where $x/4 = x - 60$). In this interval, the graph of $y = |x/4| - |x - 60|$ with the x-axis forms $\triangle ABC$, as shown in the second figure. The base of this triangle is $80 - 48 = 32$, its altitude (at $x = 60$) is 15, hence its area is 240. Kite $ABCD$ is the graph of $|x - 60| + |y| = |x/4|$; the area enclosed by it is $2 \cdot 240 = 480$.

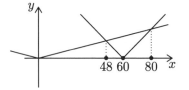
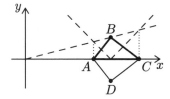

5. (588)   Rewrite the given equation in the form

$$(y^2 - 10)(3x^2 + 1) = 3 \cdot 13^2,$$

and note that, since y is an integer and $3x^2 + 1$ is a positive integer, $y^2 - 10$ must be a positive integer. Consequently, $y^2 - 10 = 1, 3, 13, 39, 169$ or 507, implying that $y^2 = 11, 13, 23, 49, 179$ or 517. Since the only perfect square in the second list is 49, it follows that $y^2 - 10 = 39$, implying that $3x^2 + 1 = 13$, $x^2 = 4$ and $3x^2y^2 = 12 \cdot 49 = 588$.

6. (193)   Since trapezoids $XYQP$ and $ZWPQ$ have the same area, and since their parallel sides are of the same length, their heights must also be equal, each of length $BC/2$. Moreover, since XY is one fourth of the perimeter of rectangle $ABCD$, it follows that $XY = (AB+BC)/2$, and that the area of trapezoid $XYQP$ is $\frac{PQ+[(AB+BC)/2]}{2} \cdot \frac{BC}{2}$. Since this must be equal to one fourth of the area of rectangle $ABCD$, one finds that

$$\frac{PQ + [(AB + BC)/2]}{2} \cdot \frac{BC}{2} = \frac{AB \cdot BC}{4},$$

from which $AB = BC + 2 \cdot PQ = 19 + 2.87 = 193$ cm.

Note. This problem arose from the following puzzle: how can one divide a rectangular cake, with a thin layer of icing on the top and sides, into n pieces of equal volume, so that each piece has the same amount of icing? The reader may wish to explore such equal-division problems for $n > 4$.

7. (070)   Since both 1000 and 2000 are of the form 2^m5^n, the numbers a, b and c must also be of this form. More specifically,

$$a = 2^{m_1}5^{n_1}, \qquad b = 2^{m_2}5^{n_2}, \qquad c = 2^{m_3}5^{n_3}, \qquad (1)$$

where the m_i and n_i are non-negative integers for $i = 1, 2, 3$. Then, in view of the definition of $[r, s]$, and since

$$[a, b] = 2^3 5^3, \qquad [b, c] = 2^4 5^3, \qquad [c, a] = 2^4 5^3, \qquad (2)$$

the following equalities must hold:

$$\max\{m_1, m_2\} = 3, \quad \max\{m_2, m_3\} = 4, \quad \max\{m_3, m_1\} = 4, \quad (3)$$

and

$$\max\{n_1, n_2\} = 3, \quad \max\{n_2, n_3\} = 3, \quad \max\{n_3, n_1\} = 3. \quad (4)$$

To satisfy (3), we must have $m_3 = 4$, and either m_1 or m_2 must be 3, while the other one can take the values of 0, 1, 2, or 3. There are 7 such ordered triples, namely (0,3,4), (1,3,4), (2,3,4), (3,0,4), (3,1,4), (3,2,4) and (3,3,4).

To satisfy (4), two of n_1, n_2 and n_3 must be 3, while the third one ranges through the values of 0, 1, 2 and 3. The number of such ordered triples is 10; they are (3,3,0), (3,3,1), (3,3,2), (3,0,3), (3,1,3), (3,2,3), (0,3,3), (1,3,3), (2,3,3) and (3,3,3).

Since the choice of (m_1, m_2, m_3) is independent of the choice of (n_1, n_2, n_3), they can be chosen in $7 \cdot 10 = 70$ different ways. This is the number of ordered triples (a, b, c) satifying the given conditions.

8. (112)   By first writing the inequalities in the form $\frac{13}{7} < \frac{n+k}{n} < \frac{15}{8}$, we can see that they are equivalent to

$$48n < 56k < 49n.$$

Consequently, the problem is to find the longest open interval $(48n, 49n)$ that contains exactly one integral multiple of 56.

Since the length of the above interval is n, it contains $n - 1$ integers. If $n - 1 \geq 2 \cdot 56$, the interval will contain at least two multiples of 56. Hence, $2 \cdot 56 = 112$ is the largest candidate for n. Indeed, we find that

$$48 \cdot 112 = 56 \cdot 96 < 56 \cdot 97 < 56 \cdot 98 = 49 \cdot 112,$$

also exhibiting that $k = 97$ is the unique positive integer corresponding to $n = 112$.

9. (033)   Noting that $\angle APB = \angle BPC = \angle CPA = 120°$, and applying the Law of Cosines to $\triangle APB$, $\triangle BPC$ and $\triangle CPA$, we find that

$$(AB)^2 = (PA)^2 + (PB)^2 + PA \cdot PB = 100 + 36 + 60 = 196 \quad (1)$$
$$(BC)^2 = (PB)^2 + (PC)^2 + PB \cdot PC = 36 + (PC)^2 + 6PC, \quad (2)$$

and

$$(CA)^2 = (PC)^2 + (PA)^2 + PC \cdot PA = (PC)^2 + 100 + 10PC. \quad (3)$$

Since $(AB)^2 + (BC)^2 = (CA)^2$ by the Pythagorean Theorem, it follows from (1), (2) and (3) that

$$196 + [36 + (PC)^2 + 6PC] = (PC)^2 + 100 + 10PC,$$

from which $PC = 33$.

Remark. More generally, if the sides of the right triangle measure a, b and c, and the distances from its "Fermat point" to its vertices are x, y, and z, as shown in the figure below (where

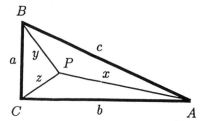

the right angle is at C rather than B), then by substituting into $a^2 + b^2 = c^2$, the system of equations

$$a^2 = y^2 + z^2 + yz, \quad b^2 = x^2 + z^2 + xz, \quad c^2 = x^2 + y^2 + xy,$$

yields

$$y = z(x + 2z)/(x - z). \quad (4)$$

The data for the present problem was chosen by letting $x = 2k+4$, $z = 2k$, and hence $y = k(3k+2)$, and then letting $k = 3$.

Alternatively, one could rewrite Eq. (4) as

$$2z^2 + (x + y)z - xy = 0,$$

and force its determinant, $x^2 + y^2 + 10xy$, to be a perfect square. The general solution of the resulting Diophantine equation yields $(x, y, z) = (m^2 - n^2, 2n(5n - m), (m - n)(5n - m)/2)$, where m and n are positive integers of the same parity and $n < m < 5n$. The present problem's data stems from the choice $(m, n) = (19, 11)$.

10. (120)   Let v_1 denote Al's speed (in steps per unit time) and t_1 his time. Similarly, let v_2 and t_2 denote Bob's speed and time. Moreover, let v be the speed of the escalator, and let x be the number of steps visible at any given time. Then, from the information given,

$$v_1 = 3v_2, \qquad v_1 t_1 = 150, \qquad v_2 t_2 = 75. \qquad (1)$$

From (1) follows that

$$t_2/t_1 = 3/2. \qquad (2)$$

We also know that $x = (v_2 + v)t_2 = (v_1 - v)t_1$, from which we have $v = (x - 75)/t_2 = (150 - x)/t_1$, and hence

$$t_2/t_1 = (x - 75)/(150 - x). \qquad (3)$$

Therefore, from (2) and (3), upon setting their right sides equal, we find that $x = 120$.

Alternative Solution. Assume that Al and Bob start at the same time from their respective ends of the escalator. Then the number of steps initially separating them is the same as the number of visible steps on the escalator. Hence, to solve the problem, we must find the number of steps each of them takes until they meet, and then add these two numbers.

Since Al can take $3 \cdot 75 = 225$ steps while Bob takes 75 steps, it follows (from $150 = (2/3) \cdot 225$) that Al walks down the escalator in $2/3$ of the time it takes Bob to walk up. Therefore, they meet $2/5$ of the way from the bottom of the escalator. To that point, Al takes $(3/5) \cdot 150 = 90$ steps, while Bob takes $(2/5) \cdot 75 = 30$ steps. As indicated above, the sum of these, 120, is the number of visible steps of the escalator.

Note. The second solution illustrates a general way to find the number of exposed steps on a moving escalator: find a friend to start simultaneously at the opposite end, and simply add the number of steps you have each taken before you meet.

Such escalator problems were rather popular at one time. This one was fashioned after a problem of Henry Dudeney (1857-1930), England's most famous creator of puzzles.

11. (486)   To solve the problem, we must find integers n and k such that n is non-negative, k is as large as possible, and

$$3^{11} = (n+1) + (n+2) + \cdots + (n+k). \tag{1}$$

Noting that

$$
\begin{aligned}
(n+1) &+ (n+2) + \cdots + (n+k) \\
&= [1 + 2 + \cdots + (n+k)] - [1 + 2 + \cdots + n] \\
&= \frac{(n+k)(n+k+1)}{2} - \frac{n(n+1)}{2} \\
&= k(k + 2n + 1)/2,
\end{aligned}
$$

it follows that (1) is equivalent to

$$k(k + 2n + 1) = 2 \cdot 3^{11}. \tag{2}$$

In solving (2), we must ensure that the smaller factor, k, is as large as possible, and that n is a non-negative integer. These conditions lead to $k = 2 \cdot 3^5 = 486$, $n = 121$ and $3^{11} = 122 + 123 + \cdots + 607$.

Alternative Solution. Let m be the average of the k consecutive integers. If k is odd, then m must be the middle integer, and $km = 3^{11}$. Now $k = 3^5$ and $m = 3^6$ is the best we can do, for if $k = 3^6$ then $m - (k-1)/2$, the smallest summand, is negative. But if k is even, then m lies halfway between the middle two integers in the sum. Thus $(2m)k = 2 \cdot 3^{11}$ and now the largest even divisor of $2 \cdot 3^{11}$ which does not give rise to a negative first summand is $2 \cdot 3^5 = 486$. This is the answer.

12. (019)   We solve the equivalent problem of finding the smallest positive integer n for which

$$n^3 + 1 < (n + 10^{-3})^3. \tag{1}$$

This is equivalent to the given problem because

$$n < \sqrt[3]{m} < n + 10^{-3} \iff n^3 < m < (n + 10^{-3})^3,$$

and because if some integer m satisfies the double inequality on the right above, then $n^3 + 1$ is the smallest such m.

Rewriting (1) in the form

$$\frac{1000}{3} < n^2 + \frac{n}{1000} + \frac{1}{3,000,000}, \qquad (2)$$

we observe that n^2 must be near $1000/3$, for the contributions of the other two terms on the right side of (2) are relatively small. Consequently, since $18^2 < 1000/3 < 19^2$, we expect that either $n = 18$ or $n = 19$. In the first case, (2) is not satisfied; this can be verified by an easy calculation. It is even easier to show that $n = 19$ satisfies (2), so it is the smallest positive integer with the desired property. The corresponding $m = 19^3 + 1 = 6860$ is the smallest positive integer whose cube root has a positive decimal part which is less than $1/1000$.

13. (931) The key property of bubble pass is that immediately after r_k is compared with its predecessor and possibly switched, the current r_k is the largest member of the set $\{r_1, r_2, \ldots, r_k\}$. Also, this set is the same at this point as it was originally, though the order of its elements in the sequence may be very different. These assertions can be verified by induction.

For a number m, which is initially before r_{30} in the sequence, to end up as r_{30}, two things must happen: m must move into the 30th position when the current $r_{29}(= m)$ and r_{30} are compared, and it must not move out of that position when compared to r_{31}. Therefore, by the key property, m must be the largest number in the original $\{r_1, r_2, \ldots, r_{30}\}$, but not the largest in the original $\{r_1, r_2, \ldots, r_{31}\}$. In other words, of the first 31 numbers originally, the largest must be r_{31} and the second largest must be m, which in our case is r_{20}.

Whatever the first 31 numbers are, there are 31! equally likely orderings of them. Of these, 29! have the largest of them in the 31st slot and the second largest in the 20th slot. (The other 29 numbers have 29! equally likely orderings.) Thus the desired probability must be $\dfrac{p}{q} = \dfrac{29!}{31!} = \dfrac{1}{930}$, and hence $p + q = 931$.

Note. Bubble pass is the first step of bubble sort, one of the simplest methods for sorting studied in computer science. Although it is relatively simple, there are all kinds of challenging probability questions associated with it.

14. (373)   Since $324 = 18^2 = 4 \cdot 3^4$, we may use the factorization

$$x^4 + 4y^4 = x^4 + 4x^2 y^2 + 4y^4 - 4x^2 y^2$$
$$= (x^2 + 2y^2)^2 - (2xy)^2$$
$$= [(x^2 + 2y^2) - 2xy][(x^2 + 2y^2) + 2xy]$$
$$= [(x^2 - 2xy + y^2) + y^2][(x^2 + 2xy + y^2) + y^2]$$
$$= [(x - y)^2 + y^2][(x + y)^2 + y^2],$$

which yields in our case

$$n^4 + 324 = [(n - 3)^2 + 9][(n + 3)^2 + 9].$$

In view of this, the given fraction can be written as

$$\frac{(7^2 + 9)(13^2 + 9)(19^2 + 9)(25^2 + 9) \cdots (55^2 + 9)(61^2 + 9)}{(1^2 + 9)(7^2 + 9)(13^2 + 9)(19^2 + 9) \cdots (49^2 + 9)(55^2 + 9)}$$

which simplifies to $\dfrac{61^2 + 9}{1^2 + 9} = \dfrac{3730}{10} = 373.$

15. (462)   Let $a = BC$, $b = AC$. We will first find the hypotenuse c of $\triangle ABC$ and the altitude h on c, because these are relatively easy to compute, and because from

$$c^2 = a^2 + b^2 \text{ and } ch = ab$$

it is easy to find $a + b$ (without first finding a and b, which is harder!).

Consider the ratios of the areas of the smaller triangles surrounding S_1 and S_2, using the fact that all five of these triangles are similar to one another and to $\triangle ABC$. To simplify the notation, let T denote both $\triangle ABC$ and its area, and similarly, let T_1, T_2, T_1', T_2' and T' denote both the triangles indicated in the figures to the right and their respective areas.

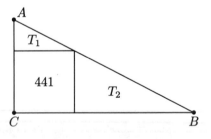

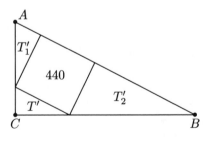

Since $\dfrac{T_1'}{T_1} = \dfrac{T_2'}{T_2} = \dfrac{440}{441}$, we find that

$$T = T_1' + T_2' + 440 + T' = \frac{440}{441}(T_1 + T_2 + 441) + T' = \frac{440}{441}T + T'.$$

Therefore, $T' = \frac{1}{441}T$, and hence the corresponding parts of triangles T' and T are in a linear ratio of 1 to 21.

It follows that $c = 21\sqrt{440}$, since the hypotenuse of T' is $\sqrt{440}$. Moreover, $h = 21h'$, where h' denotes the altitude of T' on its hypotenuse. Combining the latter equation with the observation $h = h' + \sqrt{440}$, we find that $h = \frac{21}{20}\sqrt{440}$,

$$ab = ch = (21\sqrt{440})(\frac{21}{20}\sqrt{440}) = 21^2 \cdot 22,$$

$$(a+b)^2 = c^2 + 2ab = 21^2 \cdot 440 + 2 \cdot 21^2 \cdot 22 = 21^2 \cdot 22^2,$$

and

$$AC + CB = a + b = 21 \cdot 22 = 462.$$

Note. More generally, if $\operatorname{area}(S_1) = p^2$ and $\operatorname{area}(S_2) = q^2$, one can show that

$$p = ab/(a+b) \quad \text{and} \quad q = ab\sqrt{a^2 + b^2}/(a^2 + ab + b^2). \quad (1)$$

Then, if p and q are given, either by solving the equations in (1) simultaneously (in the unknowns $a+b$ and ab) or otherwise, one can show that

$$a + b = p + (p^2/\sqrt{p^2 - q^2}) \quad \text{and} \quad ab = p(a+b). \quad (2)$$

Knowing the values of $a+b$ and ab, one can also determine the values of a and b explicitly with the help of the Quadratic Formula. In the present problem they turn out to be $21(11 \pm 3\sqrt{11})$. Since these two numbers are in the ratio of $10 + 3\sqrt{11}$ to 1, the accompanying figures are obviously not drawn to scale.

Note 2. See the related AHSME problem #21 from 1987.

1988 AIME Solutions

1. (770)   There are 2^{10} configurations of the ten buttons. We are to exclude those in which none or all of the buttons are depressed, as well as those in which exactly five buttons are depressed. Thus the total number of additional combinations is

$$2^{10} - 1 - 1 - \binom{10}{5} = 1024 - 2 - 252 = 770.$$

2. (169)   First we observe that, perhaps after a nonrepeating initial segment, the sequence $f_1(11), f_2(11), \ldots$ is periodic. To see this, it suffices to note that, for $k < 1000$,

$$f_1(k) \le f_1(999) = (9+9+9)^2 = 729 < 1000.$$

Next we compute $f_n(11)$ for the first few values of n, in the hope that the length of the period is short. This expectation is reasonable since the terms of the sequence are perfect squares, and since there are only 31 perfect squares less than 1000. We find that

$$f_1(11) = (1+1)^2 = 4,$$
$$f_2(11) = f_1(f_1(11)) = f_1(4) = 4^2 = 16,$$
$$f_3(11) = f_1(f_2(11)) = f_1(16) = (1+6)^2 = 49,$$
$$f_4(11) = f_1(f_3(11)) = f_1(49) = (4+9)^2 = 169,$$
$$f_5(11) = f_1(f_4(11)) = f_1(169) = (1+6+9)^2 = 256,$$
$$f_6(11) = f_1(f_5(11)) = f_1(256) = (2+5+6)^2 = 169.$$

We stop at this point, since $f_n(11)$ depends only on $f_{n-1}(11)$, and hence the numbers 256 and 169 will continue to alternate. More precisely, for $n \ge 4$,

$$f_n(11) = \begin{cases} 169, & \text{if } n \text{ is even,} \\ 256, & \text{if } n \text{ is odd.} \end{cases}$$

Since 1988 is even, it follows that $f_{1988}(11) = 169$.

240

3. (027)   Since

$$\log_8 x = \frac{1}{\log_x 8} = \frac{1}{3\log_x 2} = \frac{1}{3}\log_2 x$$

and

$$\log_8(\log_2 x) = \frac{1}{3}\log_2(\log_2 x),$$

the given equation is equivalent to

$$\log_2(y/3) = (1/3)\log_2 y, \tag{1}$$

where $y = \log_2 x$. From (1), $\log_2(y/3)^3 = \log_2 y$, hence $(y/3)^3 = y$, i.e.,

$$y(y^2 - 27) = 0. \tag{2}$$

Since $y \neq 0$ (for otherwise, neither side of (1) would be defined), it follows from (2) that $y^2 = (\log_2 x)^2 = 27$.

4. (020)   If n is a positive integer, then

$$\sum_{k=1}^{n} |x_k| - |\sum_{k=1}^{n} x_k| \leq \sum_{k=1}^{n} |x_k| < n, \tag{1}$$

since $|\sum_{k=1}^{n} x_k| \geq 0$ and $|x_k| < 1$ for $1 \leq k \leq n$. Since it is given that

$$\sum_{k=1}^{n} |x_k| - |\sum_{k=1}^{n} x_k| = 19, \tag{2}$$

it follows that $19 < n$. Thus the answer is 20 if there is a solution to (2) with $n = 20$. One such solution is

$$x_k = \begin{cases} .95 & \text{if } k \text{ is odd,} \\ -.95 & \text{if } k \text{ is even.} \end{cases}$$

5. (634)   The divisors of 10^{99} are of the form $2^a \cdot 5^b$, where a and b are integers with $0 \leq a \leq 99$ and $0 \leq b \leq 99$. Since there are 100 choices for both a and b, 10^{99} has $100 \cdot 100$ positive integer divisors. Of these, the multiples of $10^{88} = 2^{88} \cdot 5^{88}$ must satisfy the inequalities $88 \leq a \leq 99$ and $88 \leq b \leq 99$. Thus there are 12 choices for both a and b; i.e., $12 \cdot 12$ of the $100 \cdot 100$ divisors of 10^{99} are multiples of 10^{88}. Consequently, the desired probability is $\dfrac{m}{n} = \dfrac{12 \cdot 12}{100 \cdot 100} = \dfrac{9}{625}$ and $m + n = 634$.

6. (142) Let a and b denote the numbers in two of the squares as shown in the first figure below, and compute the two neighboring entries in terms of them. Then the common difference in the third row is $b - 2a$, while in the fourth row it is $2b - a - 74$. Consequently,

$$2a + 4(b - 2a) = 186 \quad \text{and} \quad a + 2(2b - a - 74) = 103.$$

Solving these equations simultaneously, we find that $a = 13$ and $b = 66$. Therefore, the entries in the third and fourth rows, and then in the fourth column may be computed to find that the number in the square marked by the (*) is 142. For completeness, the second figure shows the rest of the entries as well.

		*		
	74			
2a	b			186
a	2b–74	103		
0				

52	82	112	142	172
39	74	109	144	179
26	66	106	146	186
13	58	103	148	193
0	50	100	150	200

7. (110) Let x be the length of the altitude from A. Then

$$\tan^{-1}(\frac{22}{7}) = \tan^{-1}(\frac{3}{x}) + \tan^{-1}(\frac{17}{x}).$$

Taking tangents of both sides, and using the formula for $\tan(\alpha + \beta)$, we obtain

$$\frac{22}{7} = \frac{(3/x) + (17/x)}{1 - (3/x)(17/x)},$$

which simplifies to

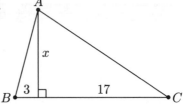

$$11x^2 - 70x - 561 = (x - 11)(11x + 51) = 0.$$

Since x must be positive, we conclude that $x = 11$, and that the area of $\triangle ABC$ is $(1/2)(3 + 17)(11) = 110$.

8. (364)   The third property is most useful in the form $f(x,z) = \dfrac{z}{z-x}f(x, z-x)$,   which is valid whenever $z > x$. To obtain this form, set $z = y+x$ in the original form, so that $y = z-x$. Using this new form of the third property and the second given property of f repeatedly, we obtain

$$f(14, 52) = \frac{52}{38}f(14, 38) = \frac{52}{38}\frac{38}{24}f(14, 24) = \frac{52}{24}\frac{24}{10}f(14, 10)$$

$$= \frac{26}{5}f(10, 14) = \frac{26}{5}\frac{14}{4}f(10, 4) = \frac{91}{5}f(4, 10)$$

$$= \frac{91}{5}\frac{10}{6}f(4, 6) = \frac{91}{3}\frac{6}{2}f(4, 2) = 91f(2, 4)$$

$$= 91\frac{4}{2}f(2, 2) = 364,$$

where the last equality is a consequence of the first given property of f.

Note. The computations may be somewhat simplified if we introduce a new function g by letting $g(x,y) = \frac{1}{xy}f(x,y)$. The student should verify that $f(x,y)$ is the least common multiple of x and y, while $g(x,y)$ is the reciprocal of the greatest common divisor of x and y. In fact, computing with g is like using Euclid's Algorithm in slow motion!

9. (192)   If the cube of an integer ends in 8, then the integer itself must end in 2; i.e., must be of the form $10k + 2$. Therefore

$$n^3 = (10k+2)^3 = 1000k^3 + 600k^2 + 120k + 8, \qquad (1)$$

where the penultimate term, $120k$, determines the penultimate digit of n^3, which must also be 8. In view of this, $12k$ must also end in 8; i.e., k must end in 4 or 9, and hence be of the form $5m + 4$. Thus

$$n^3 = (10(5m+4) + 2)^3$$
$$= 125000m^3 + 315000m^2 + 264600m + 74088. \qquad (2)$$

Since the first two terms on the right of (2) end in 000, while the last term ends in 088, it follows that $264600m$ must end in 800. The smallest m which will ensure this is $m = 3$, implying that $k = 5 \cdot 3 + 4 = 19$, and $n = 10 \cdot 19 + 2 = 192$. (Indeed, $192^3 = 7,077,888$.)

10. (840) Let V, E, D and I denote the number of vertices, edges, facial diagonals and interior diagonals of the polyhedron. We will evaluate them in succession, starting with V.

Since at each vertex of the polyhedron there is exactly one of each kind of face, and since the 4 vertices of each of the square faces must be at different vertices of the polyhedron,

$$V = 4 \times \text{number of square faces} = 4 \times 12 = 48.$$

(Note that the same value results if we consider the hexagonal or octagonal faces.)

Since there are 3 edges emanating from each vertex of the polyhedron, and since $3V$ counts each edge twice,

$$E = \frac{1}{2} \times 3V = 72.$$

To find D, note that each square face has 2 diagonals, each hexagonal face has 9 diagonals, and each octagonal face has 20 diagonals. Consequently,

$$D = 12 \times 2 \ + \ 8 \times 9 \ + \ 6 \times 20 = 216.$$

Finally, because the polyhedron is convex, we may find I by subtracting the number of edges and the number of facial diagonals from the total number of ways to connect pairs of vertices for the polyhedron. Thus

$$I = \binom{48}{2} - E - D = 1128 - 72 - 216 = 840.$$

Note. The polyhedron in question is the "great rhombicuboctahedron", also known as the (rhombi)truncated cuboctahedron. It is one of the thirteen semi-regular (Archimedean) solids. See Fejes Tóth's "Regular Figures" (Pergamon Press, 1964) or Coxeter's "Regular Polytopes" (Dover, 1973).

11. (163)   Let $y = mx + b$ be a mean line for the complex numbers $w_k = u_k + iv_k$, where u_k and v_k are real, and $k = 1, 2, \ldots, n$. Assume that the complex numbers $z_k = x_k + iy_k$, where x_k and y_k are real, are chosen on the line $y = mx + b$ so that

$$\sum_{k=1}^{n}(z_k - w_k) = 0.$$

Then

$$\sum x_k = \sum u_k, \quad \sum y_k = \sum v_k,$$

where $\sum$ means summation as k ranges from 1 to n, and $y_k = mx_k + b$, for $1 \le k \le n$. Consequently,

$$\sum v_k = \sum y_k = \sum (mx_k + b) = m \sum x_k + nb = (\sum u_k)m + nb.$$

Since in our case, $n = 5$, $b = 3$, $\sum u_k = 3$, and $\sum v_k = 504$, it follows that $504 = 3m + 15$ and hence $m = 163$.

Note. We have shown only that $m = 163$ is necessary for $y = mx + 3$ to be a mean line for the given set of points. The reader should find corresponding $z_1, z_2, \ldots, z_5$ to verify the sufficiency of this choice for m.

The definition of mean line given in this problem is not the standard one; i.e., usually it is also required that each of the segments connecting w_k to z_k be perpendicular to the mean line. Can the reader show that the two definitions are equivalent, and that a mean line for a set of complex numbers is nothing more than a line through the centroid of the set?

Query. What if $\sum u_k = 0$?

12. (441)   First observe that

(1) $\dfrac{d}{d + a} = \dfrac{\text{area}(\triangle BPC)}{\text{area}(\triangle BAC)},$

(2) $\dfrac{d}{d + b} = \dfrac{\text{area}(\triangle CPA)}{\text{area}(\triangle CBA)},$

(3) $\dfrac{d}{d + c} = \dfrac{\text{area}(\triangle APB)}{\text{area}(\triangle ACB)}.$

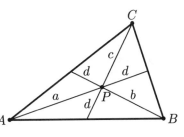

Then, since area($\triangle BPC$) + area($\triangle CPA$) + area($\triangle APB$) = area($\triangle ABC$), the sum of (1), (2) and (3) simplifies to

$$\frac{d}{d+a} + \frac{d}{d+b} + \frac{d}{d+c} = 1.$$

Multiplying through by $(d+a)(d+b)(d+c)$, expanding and grouping like terms, we may write the above as

$$2d^3 + (a+b+c)d^2 - abc = 0. \tag{4}$$

From (4), in view of the given data it follows that $abc = 2 \cdot 3^3 + 43 \cdot 3^2 = 441$.

Note. One such triangle has $a = b = 21$ and $c = 1$. We "devised" this triangle by first noting that if $s > t > 0$, then $a = b = s(s - t)$, $c = 2t^2$, and $d = t(s - t)$ satisfy the equation

$$2d^3 + (a + b + c)d^2 - abc = 0.$$

Hence, upon solving $a + b + c = 2(s^2 + t^2 - st) = 43$ and $d = st - t^2 = 3$ for s and t, one can determine a, b, and c.

Alternative Solution. Another derivation of (4) can be based on the diagram below, by first proving (by clever applications of the theorems of Menelaus and Ceva) that

$$\frac{x}{y} = \frac{r}{s} + \frac{t}{u}$$

in the figure on the left. Then it follows, in the notation of the figure on the right, that

$$\frac{a}{d} = \frac{s_1}{s_2} + \frac{s_6}{s_5}, \qquad \frac{b}{d} = \frac{s_2}{s_1} + \frac{s_3}{s_4}, \qquad \frac{c}{d} = \frac{s_5}{s_6} + \frac{s_4}{s_3}.$$

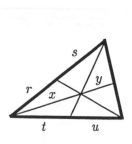
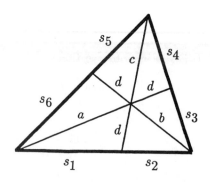

Upon computing the sum and product of these, and simplifying the resulting expressions via Ceva's Theorem, one obtains Equation (4). This solution was suggested by Sam Baethge, a high school teacher and superb geometer from Austin, Texas.

Note. For statements and proofs of the theorems of Menelaus and Ceva, see H. S. M. Coxeter & S. L. Greitzer, "Geometry Revisited," New Mathematical Library, Vol. 19.

13. (987)   Since the roots of $x^2 - x - 1 = 0$ are $p = \frac{1}{2}(1 + \sqrt{5})$ and $q = \frac{1}{2}(1 - \sqrt{5})$, these must also be roots of $ax^{17} + bx^{16} + 1 = 0$. Thus

$$ap^{17} + bp^{16} = -1 \quad \text{and} \quad aq^{17} + bq^{16} = -1.$$

Multiplying the first of these equations by q^{16}, the second one by p^{16}, and using the fact that $pq = -1$, we find that

$$ap + b = -q^{16} \quad \text{and} \quad aq + b = -p^{16}.$$

Thus

$$a = \frac{p^{16} - q^{16}}{p - q} = (p^8 + q^8)(p^4 + q^4)(p^2 + q^2)(p + q). \qquad (1)$$

Since

$$p + q = 1,$$
$$p^2 + q^2 = (p + q)^2 - 2pq = 1 + 2 = 3,$$
$$p^4 + q^4 = (p^2 + q^2)^2 - 2(pq)^2 = 9 - 2 = 7,$$
$$p^8 + q^8 = (p^4 + q^4)^2 - 2(pq)^4 = 49 - 2 = 47,$$

substitution into (1) yields

$$a = (47)(7)(3)(1) = 987.$$

Note. Substituting for p and q into (1) gives

$$a = \frac{1}{\sqrt{5}} \left[\left(\frac{1 + \sqrt{5}}{2} \right)^{16} - \left(\frac{1 - \sqrt{5}}{2} \right)^{16} \right],$$

whose right side may be recognized as (the Binet form of) the 16th Fibonacci number F_{16}. It is easiest to compute F_{16} by iteration, using the recursive definition: $F_1 = F_2 = 1$, $F_n = F_{n-1} + F_{n-2}$ for $n > 2$.

Alternative Solution. The other factor is of degree 15 and we write (with slight malice aforethought)

$$(-1 - x + x^2)(-c_0 + c_1 x - \cdots + c_{15} x^{15}) = 1 + b x^{16} + a x^{17}.$$

Comparing coefficients:

$$x^0 : c_0 = 1,$$
$$x^1 : c_0 - c_1 = 0 \implies c_1 = 1,$$
$$x^2 : -c_0 - c_1 + c_2 = 0 \implies c_2 = 2,$$

and for $3 \le k \le 15$,

$$x^k : -c_{k-2} - c_{k-1} + c_k = 0.$$

It follows that for $k \le 15$, $c_k = F_{k+1}$, the $(k+1)$st Fibonacci number. Thus $a = c_{15} = F_{16} = 987$.

Note. See the related AHSME problem #15 from 1988.

14. (084)   Let $P(u, v)$ be any point on C, and let $P^*(x, y)$ be the corresponding point on C^*, that is, the reflection of P in the line $y = 2x$. Connect P and P^* by a straight line segment as shown in the figure. Then the problem is to find an equation relating x and y.

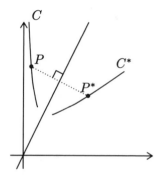

Since PP^* is perpendicular to the line $y = 2x$, its slope is $-\frac{1}{2}$. Thus

$$\frac{y - v}{x - u} = -\frac{1}{2}. \tag{1}$$

Furthermore, since the midpoint of PP^* is on the line $y = 2x$, its coordinates must satisfy the equation of this line; that is,

$$\frac{y + v}{2} = 2 \cdot \frac{x + u}{2}. \tag{2}$$

Solving (1) and (2) simultaneously for u and v yields $u = (4y - 3x)/5$ and $v = (4x + 3y)/5$. Substituting these into $uv = 1$ (which holds since P is on C) we find that

$$12x^2 - 7xy - 12y^2 + 25 = 0.$$

This is the equation of C^* in the form desired; in it $bc = (-7)(-12) = 84$.

15. (704)   At any given time, the letters in the box are in decreasing order from top to bottom. Thus the sequence of letters in the box is uniquely determined by the set of letters in the box. We have two cases: letter 9 arrived before lunch or it did not.

Case 1. Since letter 9 arrived before lunch, no further letters will arrive, and the number of possible orders is simply the number of subsets of $T = \{1, 2, \ldots, 6, 7, 9\}$ which might still be in the box. In fact, each subset of T is possible, because the secretary might have typed letters not in the subset as soon as they arrived and not typed any others. Since T has 8 elements, it has $2^8 = 256$ subsets (including the empty set).

Case 2. Since letter 9 didn't arrive before lunch, the question is: where can it be inserted in the typing order? Any position is possible for each subset of $U = \{1, 2, \ldots, 6, 7\}$ which might have been left in the box during lunch (in descending order). For instance, if the letters in the box during lunch are $6, 3, 2$, then the typing order $6, 3, 9, 2$ would occur if the boss would deliver letter 9 just after letter 3 was typed. There would seem to be $k + 1$ places at which letter 9 could be inserted into a sequence of k letters. However, if letter 9 is inserted at the beginning of the sequence (i.e., at the top of the pile, so it arrives before any after-lunch typing is done), then we are duplicating an ordering from Case 1. Thus if k letters are in the basket after returning from lunch, then there are k places to insert letter 9 (without duplicating any Case 1 orderings). Thus we obtain

$$\sum_{k=0}^{7} k \binom{7}{k} = 7(2^{7-1}) = 448$$

new orderings in Case 2.

Combining these cases gives $256 + 448 = 704$ possible typing orders.

Note. The reasoning in Case 2 can be extended to cover both cases by observing that in any sequence of k letters not including letter 9, there are $k + 2$ places to insert letter 9, counting the possibility of not having to insert it (i.e., if it arrived before lunch) as one of the cases. This yields

$$\sum_{k=0}^{7} (k + 2) \binom{7}{k} = 704$$

possible orderings, in agreement with the answer found previously.

Note 2. See the related AHSME problem #12 from 1987.

Solutions to Dropped AIME Problems

1. (028)   If x, y and z are numbers whose base 10 representations are (abc), (bca) and (cab), respectively, then one can easily verify that

$$10x - y = 999a, \qquad 10y - z = 999b, \qquad 10z - x = 999c. \qquad (1)$$

Since 999 is a multiple of 37, it follows from (1) that if any one of x, y or z is divisible by 37, then so are the others. Consequently, one may restrict attention to multiples of 37 of the desired form. These are: 37000, 37037, 37074, 37111, ..., 37999. Since $999 = 27 \cdot 37$, there are 28 of them.

2. (274)   Let p and q be the non-integral parts of x and y, respectively; that is, let $x = \lfloor x \rfloor + p$ and $y = \lfloor y \rfloor + q$. Substitution into the given system yields

$$\lfloor x \rfloor + 2\lfloor y \rfloor + q = 43.8$$
$$\lfloor y \rfloor + p + q = 18.4.$$

From the first of these, $q = .8$; from the second, $p + q = 1.4$, since $q > .4$. Thus $p = .6$, and the system reduces to

$$\lfloor x \rfloor + 2\lfloor y \rfloor = 43$$
$$\lfloor y \rfloor = 17.$$

It follows that $\lfloor y \rfloor = 17$ and $\lfloor x \rfloor = 9$, hence $x = 9.6$ and $y = 17.8$, resulting in $10(x + y) = 274$.

3. (734)   The divisibility conditions are equivalent to

$$a + c \equiv b, \qquad d + f \equiv e, \qquad g + i \equiv h,$$
$$a + g \equiv d, \qquad b + h \equiv e, \qquad c + i \equiv f,$$
$$a + i \equiv e,$$

where $\equiv$ denotes congruence modulo 11. Since $1 + 2 + 3 + \cdots + 9 = 45 \equiv 1$, and since

$$(a + c) + (g + i) + (d + f) + (b + h) + e \equiv b + h + e + e + e \equiv 4e,$$

it follows that $4e \equiv 1$, and hence $e = 3$. Therefore, from $d + f \equiv b + h \equiv a + i \equiv 3$, we have

$$\{\{d, f\}, \{b, h\}, \{a, i\}\} = \{\{1, 2\}, \{5, 9\}, \{6, 8\}\},$$

which implies that $\{c, g\} = \{4, 7\}$.

250

To maximize (ceg), we should choose $c = 7$ and $g = 4$. This is indeed possible, since the array shown on the right does satisfy the conditions of the problem. Thus, the maximum value of (ceg) is 734.

1	8	7
5	3	9
4	6	2

4. (396) Since $\triangle ABC$ and $\triangle ABP$ have the same base AB, it suffices to find the ratio of their altitudes h and h_1 on this base. Let h_2 be the altitude of $\triangle DEP$ on DE. Since $EDP \sim ABP$ with areas 25 and 36, their ratio of similitude is 5 to 6. Furthermore, $\triangle DEC$ and $\triangle ABC$ are similar with the same ratio. Thus

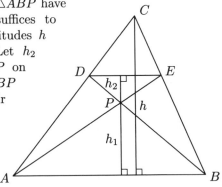

$$h_1 + h_2 = \frac{1}{6}h,$$

$$h_1 = \frac{6}{11}(h_1 + h_2) = \frac{1}{11}h.$$

Thus area$(\triangle ABC) = 11\,\text{area}(\triangle ABP) = 396$.

5. (133) Adding and subtracting the given equations, we find that

$$x^3 + y^3 = 16(x + y)$$
$$x^3 - y^3 = 10(x - y),$$

which simplify to

$$x^2 - xy + y^2 = 16$$
$$x^2 + xy + y^2 = 10,$$

since $|x| \neq |y|$. From the last two equations it follows that $x^2 + y^2 = 13$ and $xy = -3$. Consequently,

$$(x^2 - y^2)^2 = (x^2 + y^2)^2 - 4(xy)^2 = (13)^2 - 4(-3)^2 = 133.$$

Alternative Solution. Since $|x| \neq |y|$, neither x nor y is 0. Therefore, if the first equation is divided by x and the second one by y, then

$$x^2 = 13 + (3y/x),$$
$$y^2 = 13 + (3x/y). \qquad (1)$$

Thus $x^2 - y^2 = 3(y^2 - x^2)/xy$, implying $xy = -3$. We then have $y = -3/x$ and $x = -3/y$. Substituting these into (1) gives

$$x^4 - 13x^2 + 9 = 0$$
$$y^4 - 13y^2 + 9 = 0.$$

Since $|x| \neq |y|$, x^2 and y^2 are the two roots of $r^2 - 13r + 9$. Hence $\{x^2, y^2\} = \{ (13 + \sqrt{133})/2, \ (13 - \sqrt{133})/2 \}$, and $(x^2 - y^2)^2 = 133$.

6. (48)   For an arbitrary arrangement of the numbers, divide the circle into four sectors, S_1, S_2, S_3, and S_4, so that S_1 contains only the number 1, while each of the others contains three distinct consecutively placed numbers. Then the sum of the numbers in S_2, S_3 and S_4 is found to be $4 + 7 + 10 + \cdots + 28$ or 144, so the average value of the sum of the numbers in each of the sectors S_2, S_3 and S_4 is 144/3 or 48. This implies that the answer to the problem is at least 48. The fact that 48 is sufficiently large is demonstrated by exhibiting an arrangement in which this value of N is attained. Two such arrangements are shown below for the more general sequence, $k, k + d, k + 2d, \ldots, k + 9d$. In our case, $k = 1$ and $d = 3$.

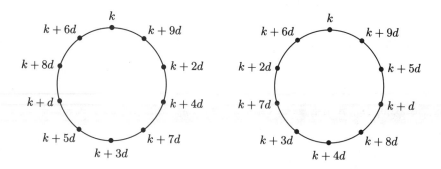

7. (172)   Upon subtracting the first from the second equation, the result may be written in the form

$$(x + y + z)(z - y) = 5^3. \qquad (1)$$

Since $x + y + z$ is positive and greater than $z - y$, it follows from (1) that $z - y = 1$ or 5, i.e. $z = y + 1$ or $z = y + 5$.

In the first case, $x + y + z = 125$. Hence $x = 125 - y - z = 125 - y - (y + 1)$, i.e. $x = 124 - 2y$, and substitution for x and z into either one of the original equations yields

$$3y^2 - 122y - 21 = 0. \tag{2}$$

By computing the discriminant of (2) or otherwise, it is easy to see that (2) has no solution in integers.

In the second case, $x + y + z = 25$, hence $x = 25 - y - z = 25 - y - (y + 5)$, i.e. $x = 20 - 2y$, and the original equations may be written in the form

$$3y^2 - 10y + 3 = 0, \tag{3}$$

whose left member can be factored as $(3y - 1)(y - 3)$. Since y is a positive integer, it follows that $y = 3$, hence $z = y + 5 = 8$ and $x = 20 - 2y = 14$. Consequently, $x^2 - yz = 14^2 - 3 \cdot 8 = 172$.

8. (540) Since AB is a diameter, $DB \perp AC$ and $AE \perp BC$. Thus we have the adjoining figure and the equations

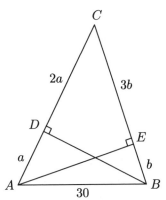

$$(CB)^2 = (4b)^2 = (2a)^2 + (DB)^2$$
$$= (2a)^2 + 30^2 - a^2$$
$$= 3a^2 + 30^2,$$
$$(AC)^2 = (3a)^2 = (3b)^2 + (AE)^2$$
$$= (3b)^2 + 30^2 - b^2$$
$$= 8b^2 + 30^2.$$

From these facts one finds that $a = 6\sqrt{5}$ and $DB = 12\sqrt{5}$. Therefore, the area of $\triangle ABC$ is $\frac{1}{2}(3a)(DB) = 540$.

9. (170) If a, b and c are digits, then

$$(abc) + (cba) = (100a + 10b + c) + (100c + 10b + a)$$
$$= 101(a + c) + 20b$$

Let $a + c = s$. If neither a nor c can equal 0, then $a + c$ can take on any integer value from 2 through 18. Thus there are 17 possible values for s. Since there are no restrictions on the digit b, there are 10 possible values for b. With these restrictions we see that the expression $101s + 20b$ can take on at most $17 \cdot 10 = 170$ different values. We show that these 170 combinations of s and b give distinct results.

Suppose s and s' are positive integers with $2 \leq s, s' \leq 18$ and that b and b' are integers with $0 \leq b,\ b' \leq 9$. If $101s + 20b = 101s' + 20b'$, then $20(b - b') = 101(s' - s)$. This last equation implies that 101 is a divisor of $20(b - b')$. Since 101 is prime and $|b - b'|$ is at most 9, we must have $b - b' = 0$. Thus $b = b'$ and $s' = s$, and indeed 170 positive integers have the reversed-sum property.

10. (463) Let θ denote the common measure PAB, PBC, and PCA, and let x, y, and z denote the lengths PA, PB, and PC, respectively. The law of cosines applied to the triangles PAB, PBC, and PCA yields

$$y^2 = x^2 + c^2 - 2xc\cos\theta,$$
$$z^2 = y^2 + a^2 - 2ya\cos\theta,$$
$$x^2 = z^2 + b^2 - 2zb\cos\theta.$$

Adding these equations produces the equation

$$2\cos\theta\,(xc + ya + zb) = a^2 + b^2 + c^2.$$

Since $\frac{1}{2}\sin\theta(xc + ya + zb)$ is the combined area of triangles PAB, PBC, and PCA, the preceding equation can be rewritten

$$\tan\theta = 4 \cdot \text{area}(ABC)/(a^2 + b^2 + c^2).$$

Now, area$(ABC) = 84$, as can be shown either via Heron's formula or by dividing the triangle into two right triangles as shown to the right. Thus

$$\tan\theta = \frac{4 \cdot 84}{590} = \frac{168}{295}.$$

Hence the answer to the problem is $168 + 295 = 463$.

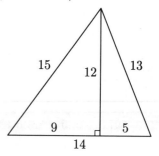

Note. The point P is known as one of the two *Brocard points* of $\triangle ABC$. The other one, Q, satisfies $\angle QAC = \angle QCB = \angle QBA$.

11. (993) Note that if n is fortunate, then so is $n + 7$, provided no carry is needed in performing the addition. We will exploit this fact by carefully choosing the digits of the numbers under consideration. In what follows, n is assumed to be superfortunate.

Let u be the units digit of n. If $u < 3$, then $u + 7 < 10$, hence $n+7$ differs from n only in its last digit. So $n+7$ is also fortunate. This is contrary to the assumption that n is superfortunate. If $u > 3$, consider the seven numbers

$$n + (10-u),\ n + (10-u) + 1, \ldots,\ n + (10-u) + 6.$$

Since n is of the form $10k + u$, these numbers differ from one another only in their units digits, which range from 0 to 6. Consequently, the sums of their digits must also differ modulo 7, and so one of them must be fortunate. Since they differ from n, and since $[n + (10-u) + 6] - n < 13$ (since $u > 3$), again we contradict that n is superfortunate. It follows that for all superfortunate numbers $u = 3$.

Since 3 itself is not even fortunate, we next consider the tens digit, t, of n. If $t < 9$, then the sum of the digits of $n + 9$ is also a multiple of 7, since by adding 9 to n, one decreases the units digit by 1 and balances this by increasing the tens digit by 1. Thus, our next candidate for a superfortunate number is 93, but once again we find that it isn't even fortunate.

Similarly, if the hundreds digit, h, of $n = \ldots 93$ is less than 9, then $n + 11$ is also fortunate, since the sum of its digits is 7 less than that of n (due to an increase of h and u by 1 each, and a decrease of t by 9). Consequently, $h = 9$, and $n = 993$ is our smallest candidate of a superfortunate number. By checking its digit sum and the digit sums of 994, 995, $\ldots$, 1005, one can readily verify that 993 is indeed superfortunate.

Note: The reader may wish to verify that the difference of 13 between consecutive fortunate numbers is indeed maximal, and that in general, superfortunate numbers are of the form $A|B|3$, where B is a block of k 9's, with $k \equiv 2 \pmod 7$, and A is an integer whose digit sum is divisible by 7; the bars denote juxtaposition.

12. (969) To exhibit a subset S with the desired properties, observe that $1000 < 32^2$, hence the subset $S = \{32, 33, 34, \ldots, 1000\}$ fulfills the conditions of the problem and has $1000 - 32 + 1$ or 969 elements.

To show that this is the largest possible size for S, let $a_i = 31 - i$, $b_i = 32 + i$, and $c_i = a_i b_i = 992 - i - i^2$, for $i = 0, 1, 2, \ldots, 29$. Then $2 = a_{29} < a_{28} < \cdots < a_0 < b_0 < b_1 < \cdots < b_{29} < c_{29} < \cdots < c_0 = 992$. Of these $3 \cdot 30$ or 90 numbers, one can choose at most $2 \cdot 30$ or 60, since of the three numbers a_i, b_i, c_i, at most two can be in S. Of the remaining $1000 - 90$ or 910 numbers, at most 909 can be chosen, since 1 is among them and its choice would exclude all others. Hence S contains at most $60 + 909$ or 969 members.

13. (265) The problem would be easier if only the columns were required to have at least one black square. Then the answer would be 7^3, because for each column each of the $2^3 = 8$ ways to color it would be acceptable, except for the one all-white coloring of that column.

However, rows have restrictions too, so 7^3 overcounts. For each of the three rows, we must subtract off the number of configurations among the 7^3 with that row all white. There are 3^3 such configurations, because, for each column, 3 of the 4 ways to color the intersection of that column with the other rows has at least one black. Subtracting these off for each row, we get $7^3 - 3 \cdot 3^3$.

But now those configurations among the 7^3 with two rows all white have been subtracted off twice, so we must add them back once. (There are no configurations among the 7^3 with all rows white, since each configuration counted has a black square in each column.) If two rows are all white, the third must be all black, and there are 3 ways to pick two rows to be all white. Thus the correct answer is

$$7^3 - 3 \cdot 3^3 + 3 = 265.$$

Note: Without saying so, we have used the Inclusion-Exclusion Principle. More generally, this method can be used to show that for an $n \times n$ board the number of admissible configurations is

$$\sum_{k=0}^{n-1} (-1)^k \binom{n}{k} (2^{n-k} - 1)^n.$$

14. (506)   Let N and N' denote the number of cyclic 3-sets and the number of 3-sets which are not cyclic, respectively. Then $N + N' = \binom{23}{3}$, and the problem is equivalent to finding the minimum value of N'.

To accomplish this, first observe that in any non-cyclic 3-set there must be one team which beats the other two. (To verify this claim, draw some triangles with arrows on the edges; teams are vertices, and the vertex at the tail beats the vertex at the head.) Conversely, for each unordered pair of teams j and k which team i beats, $\{i, j, k\}$ is a non-cyclic 3-set. Therefore, if for $i = 1, 2, \ldots, 23$, s_i denotes the number of teams beaten by team i, then in any tournament with 23 teams,

$$N' = \sum_{i=1}^{23} \binom{s_i}{2}.$$

It will be shown that N' is minimized if each s_i is equal to 11, the average value of the s_i's; this will allow us to compute N' and then N. We will also need to show that such a tournament is indeed possible.

To proceed with the above, first observe that each s_i is integral, and

   (1)   for each $i = 1, 2, \ldots, 23$, $0 \leq s_i \leq 22$;

   (2) $\displaystyle\sum_{i=1}^{23} s_i = \binom{23}{2}.$

Note that (2) holds because each of the $\binom{23}{2}$ matches counts as a win for exactly one team. Also note that from (2) it follows that the average value of s_i is $\frac{1}{23}\binom{23}{2} = 11$.

To show that N' is minimized when $s_i = 11$ for $i = 1, 2, \ldots, 23$, assume that some other sequence, $s_1, s_2, \ldots, s_{23}$, satisfying (1) and (2), leads to minimal N'. Then for some k, $s_k > 11$ and for some j, $s_j < 11$. Change s_k to $s_k - 1$, s_j to $s_j + 1$, leaving the other s_i's intact. Note that the new sequence still satisfies (1) and (2), but since $s_j + 1 \leq 11 < s_k$,

$$\binom{s_k - 1}{2} + \binom{s_j + 1}{2} = \binom{s_k}{2} - (s_k - 1) + \binom{s_j}{2} + s_j < \binom{s_k}{2} + \binom{s_j}{2},$$

and hence this new sequence leads to a smaller N' than the one assumed to be minimal. It follows that among all sequences satisfying (1) and (2), the minimal N' is obtained only if $s_i = 11$ for each i. In that case,

$$N = \binom{23}{3} - N' = \binom{23}{3} - 23\binom{11}{2} = 506.$$

It remains to be shown that such an outcome is possible in a tournament. To see this, seat all 23 captains around a round table and fix all the matches so that team i beats team j if and only if the captain of j is seated at one of the first 11 seats clockwise from the captain of i. This is a well-defined arrangement, leading to $s_i = 11$ for each $i = 1, 2, \ldots 23$. The key is that $23 > 2 \cdot 11$, so no two captains are each within 11 seats counter-clockwise of the other.

Note: In the proof of minimality above, we considered all sequences satisfying (1) and (2), not just the possible "score sequences" arising from tournaments, and found that even in this larger set, the sequence of 11's yields minimal N'. It is possible to restrict attention to possible score sequences; however, in that case one must show that if for some k, $s_k > 11$, then among the s_k teams beaten by team k, there is a team j such that $s_j < 11$.

15. (799) The procedure for converting from reverse to normal is probably familiar to anyone who has cleared lots of parentheses: a sign is switched if and only if there are an odd number of minuses to its left. The procedure for converting from normal to reverse is probably less familiar. Consider the sequence of *signs* in the series as a row of +blocks and −blocks. (For instance, $- - + + + -$ corresponds to $a_1 - a_2 - a_3 + a_4 + a_5 + a_6 - a_7$. Note that there are three numbers corresponding to the first block.) To convert from normal to reverse, keep the left-most sign of the first block the same, change the left-most sign in each of the other blocks to a minus, and change all other signs to pluses. For example, the sequence of signs given below (where spaces are shown between the blocks for easier reading)

$$+ + + +\quad - - -\quad + + +\quad - - - - -\quad + +\quad - - -$$

becomes

$$R : + + + +\quad - + +\quad - + +\quad - + + + +\quad - +\quad - + +$$

in the conversion from normal to reverse. Both rules are best discovered by working examples and looking for patterns, but formal proofs can be given by induction on the number of signs.

Now consider the normal series $a_1 \pm a_2 \pm \ldots \pm a_{1000}$ with 600 minuses and 399 pluses. Define n and m to be the maximum possible number of blocks if the first block is a $-$block and if the first block is a $+$block, respectively. Then, in view of the conversion procedure above, the answer to our problem is the larger of n and $m - 1$. However, $m \leq n + 1$, since moving an initial $+$block to the right end makes the initial block a $-$block and destroys at most one block. Thus the answer is in fact n.

When the number of blocks is maximized, it cannot be that there are both a $+$block with at least 2 members and a $-$block with at least 2 members. For then move one of the minuses between 2 of the pluses, thereby creating 2 more blocks, contrary to the assumption of their maximality.

Since there are fewer pluses, in view of the above, we may conclude that they are in 399 singleton $+$blocks, and hence there are 400 $-$blocks in the maximum configuration (starting with a $-$block). The normal series thus obtained has 799 blocks, hence the reverse series obtained from it has 799 minuses — the answer to the problem.

A Guide to the Problem Literature

In this section we will provide a selected annotated bibliography of books on problem-solving techniques, collections of (mostly) competition problems, and journals featuring active problem sections, which are available in English to students in the United States and Canada. The listings are not alphabetical; instead, we attempted to group them according to various other principles, like subject matter, national origin, publishers, etc. The ordering reflects no value judgment; in fact, some of our favorites are among the last ones listed. The selections were greatly influenced by the authors' preferences and familiarity with the literature, as well as by the mathematical level of the materials. Some of the books listed are out of print, but they should be available in better university libraries and/or through the Inter-Library Loan service. Ideally, they should all be in most high school libraries. In order to assist the readers, in case of less known publishers we provide an address either for the publisher or for a distributor of the book, if such information is known to us.

Books on Problem Solving

1. G. Pólya, *How To Solve It*, Princeton University Press, Princeton, 1973. In this pioneering work, Pólya codified most of the useful problem solving techniques and reintroduced the term "heuristic" to the literature. It is a *must* reading for anyone interested in the processes of problem solving.

2. G. Pólya, *Mathematical Discovery*, Vols. I, II, Wiley, New York, 1962, 1965; combined paperback edition, 1981. In these two volumes, Pólya elaborated on the themes of *How To Solve It*, and provided a variety of examples illustrating different problem-solving strategies.

3. G. Pólya, *Mathematics and Plausible Reasoning*, Vols. I, II, Princeton University Press, Princeton, 1954. These two volumes require a higher level of sophistication, but portions thereof should be within the grasp of capable high school students.

4. T. Butts, *Problem Solving in Mathematics*, Scott, Foresman and Company, Glenview, Illinois, 1973. A challenging little book, written in Pólya's tradition. Unfortunately, it has been out of print for some years.

261

5. S. Moses, *The Art of Problem Solving*, Transworld Publishers, London, 1974. A stimulating but little known account of the principles underlying problem solving in mathematics.

6. L. C. Larson, *Problem Solving Through Problems*, Springer-Verlag, New York, 1983. While this book is addressed mostly to upper level undergraduates, much of it can be read by experienced younger readers as well. The first chapter of this book provides further examples for the heuristics identified by Pólya.

7. T. S. C. Tao, *Solving Mathematical Problems: A Personal Perspective*, Deakin University Press (Geelong, Victoria 3217, Australia), 1992. This is a vibrant account of the thinking processes of one of the youngest and most outstanding participants of the International Mathematical Olympiads.

8. A. Soifer, *Mathematics as Problem Solving*, Center for Excellence in Mathematical Education (885 Red Mesa Drive, Colorado Springs, Colorado 80906), 1987. Interspersed with the 200 problems and their solutions, this book contains a number of suggestions for problem-solving strategies, similar to those of Pólya.

9. R. Rusczyk and S. Lehoczky, *The Art of Problem Solving*, Vols. I and II, Greater Testing Concepts (P.O. Box 5014, New York NY 10185–5014), 1994. Separate solution booklets are also available to these two volumes, authored by two of the organizers of the Mandelbrot Mathematical Competition, which should be familiar to many of the readers of this book. The third organizer of this competition, Sam Vandervelde, was a recent winner of the USAMO (along with Richard Rusczyk).

10. A. Lozansky and C. Rousseau, *Winning Solutions*, Springer-Verlag, New York, 1996. An excellent selection of olympiad and olympiad-level problems, grouped according to subject matter, with detailed introductory materials and appropriate sets of exercises.

11. *Mathematical Toolchest*, Australian Mathematical Trust (P.O. Box 1, Belconnen, ACT 2616, Australia), 1992. This little booklet provides an outline of the unofficial curriculum of the International Mathematical Olympiads, giving brief explanations, topic by topic, of the knowledge base necessary to resolve IMO level problems. Each of the nine chapters

concludes with a list of references to encourage further study.

Collections of Problems

The following books appeared in the New Mathematical Library (NML) series of The Mathematical Association of America (MAA, 1529 Eighteenth Street, NW, Washington DC 20036–1385). While we list only a few books in this series, we strongly recommend all of them. The MAA is the parent organization for the American High School Mathematics Examination (AHSME), the American Invitational Mathematics Examination (AIME), the USA Mathematical Olympiad (USAMO), and for the American participation in the International Mathematical Olympiad (IMO).

1. *Contest Problem Books* I, II, III, IV (NML Vols. 5, 17, 25, 29), MAA, Washington DC, 1961, 1966, 1973, 1983. These are the predecessors of the present volume, covering the AHSME during the years 1950–60, 1961–65, 1966–72, 1973–82, respectively.

2. *International Mathematical Olympiads, 1959–1977*, (NML Vol. 27), The Mathematical Association of America (MAA, 1978). This volume was compiled by the late Samuel L. Greitzer, longtime leader of the USA teams to the IMOs; it covers the problems and solutions of the first fifteen IMOs.

3. *International Mathematical Olympiads, 1979–1985*, (NML Vol. 31), MAA, Washington DC, 1986. This volume was compiled by Murray S. Klamkin, who served as the most outstanding trainer of the USA teams to the IMOs; it covers the problems and solutions of the next seven IMOs, along with forty additional olympiad level problems with detailed solutions. In its Appendix A the reader will find a complete description of the operation of the IMO, while Appendix B tabulates the results of the first twenty-six IMOs.

4. *USA Mathematical Olympiads, 1972–86*, (NML Vol. 33), MAA, Washington DC, 1988. This volume was also compiled by Murray S. Klamkin. As in the previous volume, the solutions are grouped according to subject matter (algebra, number theory, plane geometry, solid geometry, geometric inequalities, inequalities, and combinatorics & probability). The appendix lists the winners of the first fifteen USAMOs.

5. *Hungarian Problem Books* I, II (NML Vols. 11, 12), MAA, Washington DC, 1963. Translated by Elvira Rapaport,

these two volumes cover the problems and solutions of the Eötvös Competition during the years 1894–1905 and 1906–1928, respectively. These were originally compiled by József Kürschák; the competition is presently named after him.

6. I. M. Yaglom, *Geometric Transformations* I, II, III (NML Vols. 8, 21, 24), MAA, Washington DC, 1962, 1968, 1973. Translated from the Russian (Vols. I and II by Allen Shields and Vol. III by Abe Shenitzer), these three volumes cover translations, rotations, similarity transformations, as well as (in Vol. III) affine and projective transformations. Following brief introductions, the material is presented through a series of problems, whose detailed solutions occupy the bulk of these books.

7. R. Honsberger, *Ingenuity in Mathematics* (NML Vol. 23), MAA, Washington DC, 1970. This volume consists of 19 essays in which Honsberger develops the solution to a variety of interesting problems in a most entertaining and engaging style.

We also strongly recommend Honsberger's other books (to be listed below without further comments), along with the following other volumes in the MAA's Dolciani Mathematical Expositions (DME) series.

8. R. Honsberger, *Mathematical Gems* I, II, III (DME Vols. 1, 2, 9), MAA, Washington DC, 1973, 1976, 1985.

9. R. Honsberger, *Mathematical Morsels* (DME Vol. 3), MAA, Washington DC, 1978.

10. R. Honsberger, *More Mathematical Morsels* (DME Vol. 10), MAA Washington DC, 1991.

11. R. Honsberger, *Mathematical Plums* (DME Vol. 4), MAA, Washington DC, 1979.

12. R. Honsberger, *From Erdős to Kiev* (DME Vol. 17), MAA, Washington DC, 1996.

13. V. Klee and S. Wagon, *Unsolved Problems in Plane Geometry and Number Theory* (DME Vol. 11), MAA, Washington DC, 1991. This is a wonderful introduction to the role of unsolved problems in the development of mathematics. Most of the problems in this book are intuitively appealing and easily understood, and yet only partial answers are known to them. These are sketched in the book, along with the necessary

mathematical background, with extensive references to the relevant literature.

14. P. R. Halmos, *Problems for Mathematicians, Young and Old* (DME Vol. 12), MAA, Washington DC, 1991. This book is best described by the author himself, who said: "I wrote this book for fun, and I hope you will read it the same way. It was fun indeed — the book almost wrote itself. It consists of some of the many problems that I started saving and treasuring a long time ago." It should be noted that Paul Halmos is the author of several still very popular college text books, and was instrumental (as the editor of the series) for Springer-Verlag's excellent Problem Books in Mathematics.

15. G. T. Gilbert, M. I. Krusemeyer, and L. C. Larson, *The Wohascum County Problem Book* (DME Vol. 14), MAA, Washington DC, 1993. Many of the problems in this book were first posed as weekly challenges for undergraduates at either Carleton College or St. Olaf College, in preparation for the Putnam Examination. There is a list of prerequisites for each problem, and they are also classified by subject matter. Many of the problems should be accessible to high school students.

16. R. M. Young, *Excursions in Calculus* (DME Vol. 13), MAA, Washington DC, 1992. While this book is addressed primarily to well-trained students and teachers of calculus, much of its material can be appreciated by all students of mathematics who appreciate the elegant interplay of the continuous and discrete sides of mathematics. There are over 400 well-chosen problems complementing the six essays presented by the author.

17. J. D. E. Konhauser, D. Velleman, and S. Wagon, *Which Way did the Bicycle Go?* (DME Vol. 18), MAA, Washington DC, 1996. This book contains complete solutions to the best problems posed in 25 years of the "Problem of the Week" program, initiated by the late Professor Konhauser.

The following two books present the problems and solutions of the Putnam Examination for students at the university level. The first of these volumes contains detailed solutions and many insightful comments, while the second one is mostly a compilation of the materials previously published in the *American Mathematical Monthly* and *Mathematics Magazine*. While many of the Putnam problems are beyond the reach of most high school students, it should be noted that in recent years several winners of the USAMO were among the Putnam Fellows (top five winners) just one year after they left high school.

18. A. M. Gleason, R. E. Greenwood, and L. M. Kelly, *The William Lowell Putnam Mathematical Competition, Problems and Solutions: 1938–1964*, MAA, Washington DC, 1980.

19. G. L. Alexanderson, L. F. Klosinski, and L. C. Larson, *The William Lowell Putnam Mathematical Competition, Problems and Solutions: 1965–1984*, MAA, Washington DC, 1985.

The next two books are listed here since they were prepared primarily for students training for the Putnam Examination. They contain a section of hints prior to the presentation of detailed solutions.

20. K. Hardy and K. S. Williams, *The Green Book, 100 Practice Problems for Undergraduate Mathematics Competitions*, Integer Press (P.O. Box 6613, Station J, Ottawa, Ontario, Canada), 1985.

21. K. Hardy and K. S. Williams, *The Red Book, 100 Practice Problems for Undergraduate Mathematics Competitions*, Integer Press, Ottawa, 1988.

The next two booklets contain the problems of the Canadian Mathematical Olympiad, with detailed solutions, along with a list of the prize winners for the years indicated. It should be noted that the second volume is bilingual (English and French). The third book is listed below since originally this work was to be published as *1001 Problems in High School Mathematics* by the Canadian Mathematical Society.

22. *The First Ten Canadian Mathematics Olympiads, (1969–1978)*, Canadian Mathematical Society (577 King Edward, Suite 109, P.O. Box 450, Station A, Ottawa, Ontario K1N 6N5, Canada), 1978.

23. *The Canadian Mathematics Olympiads, (1979–1985)*, Canadian Mathematical Society, Ottawa, 1988.

24. E. J. Barbeau, M. S. Klamkin, and W. O. J. Moser, *Five Hundred Mathematical Challenges*, MAA, Washington DC, 1995. This volume is based on the preliminary edition of five booklets published in Canada. We strongly believe that

students of all levels will greatly benefit from this book.

Next we group together some excellent collections of problems translated from the Russian. Some of the problems appeared in various competitions, while others were developed to be used as building blocks for mastering subject matter. We included Soifer's two other books in this category, since they were written in the Russian tradition, along with the books on the International Mathematics Tours of the Towns, since they were first introduced in the former Soviet Union.

25. *The USSR Olympiad Problem Book*, Dover Publications, Mineola NY, 1993. This is the first volume of the original Russian edition of three volumes (compiled by D. O. Shklarsky, N. N. Chentzov, I. M. Yaglom), which was originally published by W. H. Freeman and Company in 1962. The problems of this volume are also mostly from the Moscow Mathematical Olympiads; they cover the areas of number theory and algebra. Unfortunately, the other two volumes of this book (covering the areas of plane and solid geometry) were never translated into English.

26. I. F. Sharygin, *Problems in Plane Geometry*, Mir Publishers (2 Pervy Rizhsky Pereulok, I–110, GSP, Moscow, 129820, Russia), 1988.

27. I. F. Sharygin, *Problems in Solid Geometry*, Mir Publishers, Moscow, 1986. While these two volumes bear the same titles, they are not the translations of Volumes 2 and 3 of *The USSR Olympiad Problem Book*. Nevertheless, they are nice collections, similarly based on the Moscow Mathematical Olympiads and on problems originally published in *Kvant*, the Russian periodical for high school students.

28. A. M. Yaglom and I. M. Yaglom, *Challenging Mathematical Problems with Elementary Solutions* I, II, Holden-Day, San Francisco, California, 1964, 1967. These volumes contain some of the problems (with detailed solutions) that were originally discussed at sessions of the famous School Mathematics Circle (for secondary school students) at Moscow University or were given in the associated Moscow Mathematical Olympiads.

29. E. B. Dynkin et al, *Mathematical Problems: An Anthology*, Gordon and Breach, New York, 1969. This is a collection of 143 problems, followed by a section of hints and complete

solutions, and supplemented by 35 additional problems.

30. D. Fomin and A. Kirichenko, *Leningrad Mathematical Olympiads, 1987–1991*, MathPro Press (P.O. Box 713, Westford, Massachusetts, 01886–0021), 1994. Outside of Hungary, the first olympiad-type competition was held in St. Petersburg, in 1934. This competition is unique in that in its last two rounds the students compete orally. The problems presented here are those of the last two rounds during the years indicated.

31. A. Soifer, *How Does One Cut a Triangle?*, Center for Excellence in Mathematical Education, 1990. A nice little collection of mostly geometry problems.

32. V. Boltyanski and A. Soifer, *Geometric Etudes in Combinatorial Mathematics*, Center for Excellence in Mathematical Education, 1991. A bit deeper than the previous volume, but still accessible to high school students.

33. V. Boltyansky and I. Gohberg, *Results and Problems in Combinatorial Geometry*, Cambridge University Press, 1985. While the main part of this book is suitable for high school students, it is definitely at an advanced level. At the end of the book several unsolved problems are listed.

34. *International Mathematics Tournament of the Towns, Book 1: From beginning to March 1989*, Australian Mathematics Trust, Belconnen, 1989. This booklet and the following three volumes feature the problems and solutions of an international problem-solving competition among cities around the world, in which small and large towns participate on an equal basis due to a democratic formula which takes into account the town's population. The books were edited by Peter J. Taylor; the problems and their solutions are among the most ingenious ones in the literature.

35. *The International Mathematics Tournament of the Towns, 1980–1984*, Australian Mathematics Trust, Belconnen, 1993.

36. *The International Mathematics Tournament of the Towns, 1984–1989*, Australian Mathematics Trust, Belconnen, 1992.

37. *The International Mathematics Tournament of the Towns, 1989–1993*, Australian Mathematics Trust, Belconnen, 1994.

We next list some of the contributions of the excellent Polish school of mathematics to the English problems literature.

38. S. Straszewicz, *Mathematical Problems and Puzzles from the Polish Mathematical Olympiads*, Pergamon Press, Oxford, 1965. A beautiful collection of problems, which has been out of print for several years.

39. M. E. Kuczma, *Problems: 144 Problems from the Austrian-Polish Mathematical Competition, 1978–1993*, Academic Distribution Center (1216 Walker Road, Freeland, Maryland 21053), 1993. Each of these competitions consisted of nine problems, so there are 144 problems with complete solutions and many insightful comments in this volume, authored by one of the most eminent problemists of our era. It should be noted that the books published by the Australian Mathematics Trust, the "Problems, Problems, Problems" series which features problems from the Canadian Mathematics Competitions, as well as Derek Holton's "Problem Solving Series" (from New Zealand) are also available from the Academic Distribution Center.

40. M. E. Kuczma, *40th Polish Mathematics Olympiad, 1989/1990*, SCT Publishing (AMK Central Post Office, P.O. Box 0581, Singapore 915603, Republic of Singapore), 1996.

41. W. Sierpinski, *A Selection of Problems in the Theory of Numbers*, Oxford University Press, 1964.

42. W. Sierpinski, *250 Problems in Elementary Number Theory*, Elsiviere, New York, 1970. These two volumes are most strongly recommended to all students who wish to learn number theory through problems.

43. H. Steinhaus, *Mathematical Snapshots*, Oxford University Press, 1969. A delightful collection of mathematical investigations into a variety of areas.

44. H. Steinhaus, *One Hundred Problems in Elementary Mathematics*, Basic Books, New York, 1964. A wonderful selection of truly ingenious problems.

The following books contain collections of problems from various prestigious regional competitions in the United States.

45. G. Pólya and J. Kilpatrick, *The Stanford Mathematics Problem Book*, Teachers College Press, Columbia University, New York, 1974. This volume contains the problems and solutions of the Stanford University Competitive Examination conducted by Gábor Szegő and George Pólya from 1946 to

1965. This competition was modeled after the Eötvös Competition of Hungary, which was the forerunner of all modern mathematical competitions.

46. *NYSML–ARML Contests, 1973–1983*, Mu Alpha Theta, University of Oklahoma, Norman, Oklahoma, 1984. This and the next two volumes contain the problems posed at the American (formerly, Atlantic) Regions and the New York State Mathematics League meets, since the inception of these annual competitions. The second two volumes were compiled by L. Zimmerman and G. Kessler, who served on the problems committee of ARML for many years.

47. *NYSML–ARML Contests, 1983–1988*, NCTM, Reston, Virginia, 1989.

48. *NYSML–ARML Contests, 1989–1994*, MathPro Press, Westford, Massachusetts, 1995.

49. *Saint Mary's College Mathematics Contest Problems*, Creative Publications, Palo Alto, 1972.

50. M. Saul et al, *The New York City Contest Problem Book*, Dale Seymour Publications, Palo Alto, 1985.

51. L. Fisher et al, *Brother Alfred Brousseau Problem Solving and Mathematics Competition*, Dale Seymour Publications, Palo Alto, 1985.

52. D. Barry and J. R. Lux, *The Phillips Academy Prize Examinations in Mathematics*, Dale Seymour Publications, Palo Alto, 1984.

53. J. A. Goebel, *North Carolina Mathematics League Contest Problem Book*, Dale Seymour Publications, Palo Alto, 1985.

54. G. L. Alexanderson et al, *The Santa Clara Silver Anniversary Contest Book*, Dale Seymour Publications, Palo Alto, 1985.

The next two volumes are collections of problems posed in the *Mathematics Student* from 1964 to 1973. The "plus six" refers to the fact that, in addition to the problems and solutions, the second volume also features six articles published in *Mathematics Student*.

55. M. Charosh, *Mathematical Challenges*, National Council of Teachers of Mathematics (NCTM, 1906 Association Drive, Reston, Virginia 22091–1593), 1965.

56. T. J. Hill, *Mathematical Challenges II Plus Six*, NCTM, Reston, Virginia 22091-1593, 1974.

Next we list two more volumes published by Springer-Verlag in its Problem Books in Mathematics series. The further books in the series are at a more advanced level.

57. E. J. Barbeau, *Polynomials*, Springer-Verlag, New York, 1989.

58. D. J. Newman, *A Problem Seminar*, Springer-Verlag, New York, 1982.

Our next grouping includes collections of problems on which one can hone one's algebraic, geometric, and related skills. They are excellent sources of not too difficult problems.

59. A. S. Posamentier and C. T. Salkind, *Challenging Problems in Algebra* Vols. 1, 2, The MacMillan Company, New York, 1970.

60. A. S. Posamentier and C. T. Salkind, *Challenging Problems in Geometry* Vols. 1, 2, The MacMillan Company, New York, 1970. Both of these two-volume sets are recommended, especially the ones on geometry.

61. F. Mosteller, *Fifty Challenging Problems in Probability with Solutions*, Addison-Wesley, Reading, 1965. This is one of the few books in English which is devoted entirely to problems in probability.

62. D. A. Holton and J. A. Rickard, *Problems, Problems*, Prentice-Hall of Australia, Sydney, 1980. This set of problems (with hints and solutions) comes from the Melbourne University School Mathematics Competition, sponsored by IBM Australia Limited. It is a nice collection of manageable problems.

63. M. N. Aref and W. Wernick, *Problems and Solutions in Euclidean Geometry*, Dover Publications, Mineola NY, 1968.

64. S. J. Bryant et al, *Nonroutine Problems in Algebra, Geometry, and Trigonometry*, McGraw-Hill, New York, 1965. A nice collection of mostly manageable problems.

65. D. K. Faddeev and I. S. Sominskii, *Problems in Higher Algebra*, Freeman, San Francisco, 1965.

66. V. A. Krechmar, *A Problem Book in Algebra*, Mir Publishers, 1974.

67. A. P. Hillman and G. L. Alexanderson, *Algebra through Problem Solving*, Allyn and Bacon, Boston, 1966.

Finally, we list five more books which didn't seem to fit any of the previous categories; nevertheless, they are highly recommended to our readers.

68. H. Fukagawa and D. Pedoe, *Japanese Temple Geometry Problems: Sangaku*, Charles Babbage Research Centre (P.O. Box 272, St. Norbert Postal Station, Winnipeg, R3V 1L6, Canada), 1986. Since geometry developed very differently in Japan during its 200 years of isolation, these problems — concerning the fittings and relative sizes of various inscribed figures — should be most challenging to everyone.

69. C. W. Trigg, *Mathematical Quickies*, Dover Publications, Mineola NY, 1985. A collection of 270 stimulating problems with surprisingly brief and elegant solutions. Most of the problems are within the grasp of high school students.

70. L. S. R. Mbili, *Mathematical Challenge!*, Mathematical Digest (Department of Mathematics, University of Cape Town, 7700 Rondebosch, South Africa), 1978. This is a very nice collection of "100 problems for the olympiad enthusiast", written by one of the early winners of the South African Mathematical Olympiad while still a student at the University of Cape Town.

71. A. S. Posamentier and W. Wernick, *Advanced Geometric Constructions*, Dale Seymour Publications, Palo Alto, 1988. This volume contains a wealth of nice problems with detailed solutions.

72. H. Dörrie, *100 Great Problems of Elementary Mathematics*, Dover Publications, Mineola NY, 1965. A collection of classical problems with detailed solutions and historical notes.

73. S. Vandervelde, *The First Five Years*, Greater Testing Concepts (30-B Mt. Auburn St., Cambridge MA 02138), 1996. This volume is a compilation of the team tests (with solutions) from the first five years (1990–1995) of the Mandelbrot Competition.

74. *Problems and Solutions from The Mathematical Visitor*, MathPro Press, 1996. This book contains all 344 problems (with solutions to about 250 of them) that were published in *The Mathematical Visitor* during the years 1877 to 1896. The

problems have been painstakenly sorted, edited, and reworded by Stanley Rabinowitz to conform to current mathematical style and notation.

75. *Mathematical Challenges*, Blackie & Sons Ltd. (Bishopbriggs, Glasgow G64 2NZ), 1989. This volume contains the 213 problems (with solutions) posed in the first 12 years of the competition "Mathematical Challenge", held in Scotland since 1976, under the auspices of the Scottish Mathematical Council. The problems and solutions for the next three years (1988–1991) are still in pamphlet form, but in 1994 the Scottish Mathematical Council published *Mathematical Challenges II*, covering the years 1991–1994.

Journals with Problem Sections

Since most competitions are once-a-year events, and since solving problems featured in books encourages too much isolation, it is recommended that interested students should get involved in year-round problem-solving activities. This is possible through the Mandelbrot Mathematics Competition mentioned earlier, as well as through the USA Mathematical Talent Search (USAMTS) conducted under the auspices of the Consortium for Mathematics and Its Applications (COMAP, 57 Bedford Street, Suite 210, Lexington, Massachusetts, 02173). Alternately, one should become a regular solver of problems featured in the problem sections of the following publications.

1. *The American Mathematical Monthly*, along with the next three journals listed, is an official publication of the MAA. Its Problem Section is definitely the most prestigious problem section in the world, but its problems are usually beyond the level of high school students.

2. *Mathematics Magazine* features easier problems, but most of them are still a bit too demanding even for college students. Nevertheless, one should peruse them periodically.

3. The *College Mathematics Journal* traditionally has more problems within the reach of capable high school students.

4. *Math Horizons* is primarily a publication for undergraduates, but it should appeal to advanced high school students as well. In particular, its problem section (initiated by G. Berzsenyi and presently edited by M. S. Klamkin), features

many problems suitable for students at the high school level. This publication appears four times a year.

5. *Crux Mathematicorum* is the foremost problems journal in English, with an active problem section and an informative Olympiad Corner (featuring many more problems and solutions), which was edited by M. S. Klamkin for many years. *Crux Mathematicorum* is published by the Canadian Mathematical Society (577 King Edward, Ottawa, Ontario K1N 6N5, Canada).

6. *Pi Mu Epsilon Journal* is the official publication of the Pi Mu Epsilon Mathematics Honor Society. While it is a journal for undergraduates, its problem section is not too difficult. For subscription information contact its Editor, Professor Underwood Dudley (Mathematics Department, DePauw University, Greencastle IN 46135).

7. *The Pentagon* is another semiannual publication for undergraduates; its problem section is of similar level. For subscription information contact Kappa Mu Epsilon, Division of Mathematics and Computer Science, Emporia State University, Emporia KS 66801.

8. The *Fibonacci Quarterly* is devoted mostly to investigations related to the famous Fibonaccci sequence. Its "elementary" problems cover many areas of elementary number theory and, hence, are manageable. For subscription information contact its Subscription Editor, Patty Solsaa (P.O. Box 320, Aurora SD 57002-0320).

9. *Mathematical Mayhem* is a Canadian mathematical journal for high school and university students. It was started and continues to be edited by former high school contestants. We strongly recommend it to our readers. Since January 1997, *Mathematical Mayhem* has been bundled together with *Crux Mathematicorum*.

10. *Mathematical Spectrum* is published in England for students at the high school and college level. It appears three times a year, with a lively problem section. For subscription information, contact the Applied Probability Trust (School of Mathematics, University of Sheffield, Sheffield, S3 7RH, England).

11. *Parabola* is an Australian publication, which also has a lively problem section. All correspondence should be addressed to

its Editor (School of Mathematics, University of New South Wales, P.O. Box 1, Kensington, NSW, 2033, Australia).

12. *Mathematical Digest* is a South African publication with lots of interesting articles and problems. For more information, contact its Editor (Department of Mathematics, University of Cape Town, 7700 Rondebosh, South Africa).

13. The *Journal of Recreational Mathematics* appears twice a year, featuring a Problems and a Conjectures Section. Its problems are usually of a recreational nature. It is published by Baywood Publishing Co. Inc. (26 Austin Avenue, Box 337, Amityville NY 11701).

14. *Quantum* is published six times a year. In addition to a variety of problems from its "sister journal", the Russian *Kvant*, it has a Physics Contest and a Math Investigations section, which often features problems suitable for high school students. For ordering information, contact the Journal Promotion Department of Springer–Verlag New York, Inc. (175 Fifth Avenue, New York NY 10010-7858).

15. *Consortium* is the quarterly newsletter of COMAP. In addition to its regular USAMTS column (of five problems), it has a "Problems, Puzzles, and Paradoxes" column, which is usually devoted to selections of interesting problems from various sources.

16. The *Mathematics and Informatics Quarterly* is an international publication for students interested in problem-solving. It has several active problem sections, including one in informatics (i.e., computer science), and it also features the international extension of the USAMTS (along with detailed solutions to all of its problems). To subscribe to this publication, please contact the first author, George Berzsenyi (Department of Mathematics, Rose-Hulman Institute of Technology, Terre Haute IN 47803).

17. *Mathematics Competitions* is the official journal of the World Federation of National Mathematics Competitions. Though it has no problem section, many of the articles featured in it are wonderful sources of interesting problems and problem-solving strategies. For subscription information, please contact The Australian Mathematics Trust (University of Canberra, P.O. Box 1, Belconnen, ACT 2616, Australia).

Classification of Problems

To classify these problems is not a simple task; their content is so varied and their solution possibilities so diverse that it is difficult to pigeonhole them into a few categories. Nonetheless, the following may be helpful if you wish to select a particular category of problems.

The classification is organized as follows. The main headings are Algebra, Arithmetic, Geometry, Trigonometry, and Miscellaneous (which consists mostly of what is now called Discrete Mathematics – combinatorics, algorithms, number theory, etc.) The second-level headings are usually about objects, not procedures. For instance, under algebra suppose you want problems on solving linear equations; look first for "equations, linear", not "solving". Suppose you want problems on simplifying complex-number fractions. Look first under complex numbers or fractions (again, under Algebra). (In this case "simplification" is also a heading, under "algebra, manipulation".) In the geometry section, most listings are by type of figure. So if you seek problems about tangents to circles, look under "plane figures, circles". However, there are some concepts that make sense for many types of figures, e.g., area. So if you want problems about area of polygons, look for "area, polygons" as well as "plane figures, polygons". The problems you want should be listed both places. (Indeed, most problems are listed two or three times.)

Problems are indexed by topic, not solution method. For instance, a problem about angles in a triangle may be solvable by the Law of Cosines, but there is no listing for the Law of Cosines; look under "plane figures, triangles, angles". We have chosen not to list problems by solution method because that gives away some of the challenge and it cannot be comprehensive – most problems have many solution methods, and the methods we list may not be the ones you would think of.

Numbering convention: Problem 7 from the 1985 AHSME is listed as 85-7; problem 14 from the 1983 AIME is listed as *83-14*. That is, AHSME problem numbers are printed in roman, AIME numbers in italic. Problem 13 in the Dropped AHSME section is listed as D-13 and the 8th dropped AIME problem as *D-8*. For each topic category, all the AHSME problems are listed first.

Algebra

Geometry

Trigonometry

Miscellaneous

iteration	84-11, *87-13*
calculators	84-11, D-14
codes	87-16
computers (computer language)	85-7
combinatorics (counting)	
circular arrangements	*83-7, D-6*
miscellaneous configurations	83-16, 85-9, 85-20, 87-27, D-6, *D-13*
permutations	85-11, 86-10, 87-12, 88-18, *84-11, 86-13, 87-13, 88-15*
pigeonholes	86-17
number of solutions	83-2, 83-24, 84-20, 85-21, 86-30, 87-10, 88-22, 88-30, D-8, D-10, *83-10, 84-12, 85-10, 86-13, 87-1, 87-7, 88-1, 88-10, 88-15,* D-9
subsets	*D-12*
logic	
conditional statements	86-4, 88-6
puzzles	87-17, D-16, *87-10, 88-15*
modeling	83-27, 87-13, 87-22
number theory	
common divisors and multiples	85-26, 86-24, *85-13, 87-7,* D-11
composite numbers	84-3, *84-14*
congruences	83-14
diophantine equations	84-21, 84-28, D-5, *84-10, 87-5, D-7*
divisibility and remainders	85-12, 86-4, 86-23, *83-6, 84-2, 86-5, 86-8, 87-3, 88-5, D-1, D-3*
partitions	85-18, D-15
primes	83-3, 85-12, 87-3, 87-23
squares	D-10
percents	83-7, 86-20, 87-5, 88-11, D-12, D-22